CRISIS COMMUNICATION AND THE PUBLIC HEALTH

HAMPTON PRESS COMMUNICATION SERIES
HEALTH COMMUNICATION
Gary L. Kreps, series editor

Communication in Recovery: Perspectives on Twelve-Step Groups
Lynette S. Eastland, Sandra L. Herndon, and *Jeanine R. Barr (eds.)*

Communicating in the Clinic: Negotiating Frontstage and Backstage Teamwork
Laura L. Ellingson

Media-Mediated Aids
Linda K. Fuller (ed.)

Cancer-Related Information Seeking
J. David Johnson

Changing the Culture of College Drinking: A Socially Situated Health Communication Campaign
Linda C. Lederman and *Lea P. Stewart*

Handbook of Communication and Cancer Care
Dan O'Hair, Gary L. Kreps, and *Lisa Sparks, (eds.)*

Crisis Communication and the Public Health
Matthew W. Seeger, Timothy L. Sellnow, and *Robert R. Ulmer (eds.)*

forthcoming

A Natural History of Family Cancer
Wayne Beach

Cancer Communication and Aging
Lisa Sparks, Dan O'Hair, and *Gary L. Kreps (eds.)*

Communicating Spirituality in Health Care
Maggie Wills (ed.)

CRISIS COMMUNICATION AND THE PUBLIC HEALTH

edited by

Matthew W. Seeger

Wayne State University

Timothy L. Sellnow

University of Kentucky

Robert R. Ulmer

University of Arkansas—Little Rock

HAMPTON PRESS, INC.
CRESSKILL, NJ 07626

Printed in the United States of America

Library of Congress Cataloging-in-Publication Data

Crisis communication and the public health / edited by Matthew W. Seeger, Timothy L. Sellnow, Robert R. Ulmer.

p. cm. -- (The Hampton Press communication series. Health communication)

Includes bibliographic references and indexes.

ISBN 978-1-57273-750-1 (hardbound) -- ISBN 978-1-57273-751-8 (paperbound)

1. Communication in public health. 2. Emergency management. 3. Crisis management. I. Seeger, Matthew W. (Matthew Wayne), 1957- II. Sellnow, Timothy L. (Timothy Lester), 1960- III. Ulmer, Robert R. IV. Series: Health communication (Cresskill, N.J.)

[DNLM: 1. Communication. 2. Disaster Planning. 3. Bioterrorism. 4. Disease Outbreaks. WA 295 C932 2000]

RA423.2.C75 2008
362.101'4--dc22

2007046001

Hampton Press, Inc.
23 Broadway
Cresskill, NJ 07626

Contents

Acknowledgments

The editors wish to thank the contributors for their steadfast dedication to building a safe, principled, and empathetic world through their crisis communication scholarship. In this text, we have had the honor of working with some of the most established and respected crisis communication scholars as well as some of the most promising young scholars in the field. The result is a volume that will contribute notably to our never-ending objective of being better prepared to mitigate, and manage crises of public health. True to this mission, all royalties from this book are donated to the UNICEF organization. In doing so, we support UNICEF's goal of enhancing the health, education, equality, and protection of children facing crises of all sorts throughout the world. We specifically thank the National Center for Food Protection and Defense and the Centers for Disease Control and Prevention for their support in creating opportunities to conduct many of the studies included in this volume. Our hope is that these chapters will inspire continued research and refinement of crisis communication in the public health setting.

PART I

Introduction

1

CRISIS COMMUNICATION AND THE PUBLIC HEALTH

Integrated Approaches and New Imperatives[1]

Matthew W. Seeger

Barbara Reynolds

This chapter outlines the new approaches to risk's and crisis communication that are becoming increasingly important to public health. The new environment of risks and crisis can be best understood using principles of chaos theory. Moreover, new requirements for public health communication focus on the integration of risk and crisis communication into a framework known as crisis and emergency risk communication. This chapter also describes the ways in which the public health community can learn from crisis and risks. Finally, this chapter previews the subsequent chapters and the insights they offer about crisis communication and the public health.

[1]Parts of this chapter are based on B. Reynolds and M. Seeger (2005). Crisis and emergency risk communication as an integrative model. *Journal of Health Communication*, *10*, 43–55.

The world is becoming both increasingly complex and hazardous. The emergence of new infectious diseases and reemergence of older diseases; the specter of catastrophic natural disasters such as hurricanes, earthquakes, floods, wild fires, and tsunamis; the possibility of massive technological failures, industrial disasters, transportation accidents, and toxic material spills; and the looming threat of terrorism all create complex threats of widespread loss of life, property, and security. Although a number of recent crises and disaster events stand out for consideration, the September 2001 anthrax attack was a powerful wake-up call for the public health community—even though much larger and more widespread health risks exist. The steady progression of avian influenza, the emergence of sudden acute respiratory syndrome (SARS), the upswing in skin cancer, the recent West Nile virus outbreaks, and the possibility of large and powerful hurricanes, for example, all carry much more significant public health threats than the anthrax attack.

Anthrax, however, was new in terms of the perceived threat and the intentional use of a biological agent as a terror weapon. Because anthrax was also connected in terms of time and perception to the 9/11 World Trade Center disasters, it created widespread public concern and fear. The failure of government agencies to successfully manage the threat also created public outrage. The event placed significant pressure on the public health community to communicate effectively within a context of immediacy, threat, and high uncertainty. The result was an important opportunity for learning effective crisis communication lessons. The chapters included in this volume are a response to these learning opportunities and to a renewed need for the public health community to communicate to the public about crises and risks.

Clearly, the public health community has entered a new era of heightened threats with attendant new obligations for communicating warnings and mounting effective responses. These threats are complex and often defy predication and easy interpretation. The public health community must respond in a flexible and adaptive manner and must continually learn from these events. In this chapter, we first use the propositions of chaos theory to characterize the new environment of risks and crisis and bracket the dynamic and unpredictable features of crisis. We also describe these as a consequence of the interaction of multiple, complex systems. The new requirements for public heath communication regarding risks and crisis are also described. We do so by merging traditional notions of risk communication as practiced in public health and the concept of crisis communication, more commonly associated with public relations. We also describe the ways in which the public health community can learn from crisis and risks. Institutional learning is necessary for the public health community to create new understandings, insights and capacities in response to emergent and dynamic threats. Finally, we preview the subsequent chapters and the insights they offer about crisis communication and the public health.

AN AGE OF CRISIS AND DISASTER

Crises and disasters, of course, are not new. One of pubic health's archetypal events, for example, is the influenza pandemic of 1918, when as many as 20 million people worldwide and 500,000 in the United State died. The specter of the 1918 pandemic has recently reemerged with the threat of the H5N1 strain of avian influenza. Historically, natural disasters have claimed tens of thousands of lives. Hurricanes Katrina and Rita may end up costing $300 billion in lost property, displaced millions of residents, and cost over 1,400 lives. Earthquakes generally lead the list. Several earthquakes in China and the Middle East have claimed over 200,000 lives. The July 27, 1976 earthquake in Tangshan, China, may have led to 500,000 deaths. Technological failures also have created widespread threat to public health. The nuclear accident at Chernobyl, although claiming few direct causalities, created lingering health effects for millions. The Bhopal-Union Carbide industrial accident claimed as many as 10,000 lives from direct exposure to dangerous chemicals. The ongoing health effects continue to create problems for millions of Bhopal, India, residents. In addition, crises may create widespread economic harm far beyond the scope of the immediate event. The Exxon Valdez oil spill not only affected the physical and psychological health of many Alaskans, it also contaminated many commercially important fisheries.

By their very nature, crises create severe threats to the public health and welfare, including the physical, psychological, emotional, and economic well-being of the public. In general, the more widespread and serious the threat, the more severe the crisis. Thus, an outbreak of e-coli poisoning from thousands of pounds of hamburger products distributed widely, such as the Hudson Foods case, is a more severe crisis than a local outbreak of Shigelella. The former is often associated with severe health problems and even death, whereas the latter is largely a matter of extreme discomfort. In addition, researchers have noted that a number of other factors are associated with the perceived seriousness of a threat. These include the perceived fairness of the risk, the degree of familiarity, and whether the risk is voluntary among other factors (Sandman, 1993). In addition, preexisting perceptions are important to the perceived level of risk.

Chaos and Public Health Crises

One important factor in understanding the increased risk the public faces is the chaotic nature of disease systems, particularly as they interact with other systems. Crises and disasters are increasingly dynamic events, interacting with technological, social, political, economic, and natural factors in highly complex, unpredictable and unanticipated ways. One of the themes of modern public health is the emergence of complex, unanticipated risks. Probably one of the best examples described in subsequent chapters is SARS. The emergence and spread of SARS was a consequence of social practices, demographic and travel patterns, and political systems. The SARS episode, for example, included the interaction of a number of complex systems in unforeseen ways. The emergence of the disease in parts of mainland China was quickly exacerbated by population demographics and international travel patterns. Many residents of Hong Kong relocated to Toronto in the late 1990s to avoid the possibility of government restrictions and still commute by air to Hong Kong. The SARS outbreak in Toronto was a function of this international travel pattern. Moreover, the SARS virus was actually spread via the public health system. A significant proportion of the second generation of cases was traced to infected patients visiting emergency rooms or being transported in ambulances. In fact, a number of emergency medicine workers became infected with SARS.

SARS is illustrative of the ways in which unexpected factors can interact in chaotic ways. Similarly, the cases of hepatitis in green onions, associated with Chi-Chi's restaurants and the emergence of West Nile virus, also discussed in following chapters, are rooted in complex, unanticipated interactions.

Most kinds of public health crises can be described as severe threats to the physical and psychological security, stability, health, and well-being of the public resulting from complex, nonlinear, and unanticipated interactions. As described earlier, these interactions often involve social, economic, political, and technological forces as well as natural disease or other natural processes. The developing threat of avian influenza (H5N1) involves not only the natural disease process, but also global economic trends leading to widespread commercial poultry production in southeast Asia. These economic factors have interacted with wild birds' natural migration patterns. Crisis theorists have used the concepts of chaos theory (CT) to emphasize these complex interactions (see Perrow, 1983; Seeger, Sellnow, & Ulmer, 2003). CT argues that systems, particularly complex ones, may experience radical breakdowns or disruptions, or what is described as *bifurcations*. These breakdowns are a result of the inherent complexity of the system and its interaction with other systems. CT also emphasizes the role of small variances in the system leading to disproportional impacts and breakdown. This concept, sometimes described as the *butterfly effect*, suggests that changes that might be dismissed as irrelevant and insignificant can have widespread, often devastating effects. Relatively minor changes in agricultural practices, or minor mistakes in food processing, for example, have led to widespread contamination and disease systems. Although chaos theory is a pessimistic view of the world, it also describes ways in which systems such as communities and organizations emerge from crises.

Public health sits at the intersection between disease systems and processes and dynamic social systems, structures, and institutions. These conditions are inherently complex and dynamic. In fact, epidemiology is filled with examples of puzzling and unanticipated routes of exposure, infection, and cross-contamination. It often falls on public health to sort out these complexities and communicate appropriate guidelines and advice to the public.

According to CT, inherent to any system is an underlying order or stability, usually created by some natural, albeit, not obvious, patterns or set of forces. Following a major bifurcation or breakdown, these forces and patterns reassert themselves, helping the system recover and reconstitute itself. This self-organization begins immediately as those affected by a crisis and those working to manage it assess what needs to be done to limit the harm and recover. The patterns of relationships in communities and the structures of community groups and nongovernmental organizations often begin immediately to mitigate harm and rebuild. One way in which self-organization is initiated and facilitated is through effective communication.

Communication and Public Health Crises

In general, the management of natural disasters and public health emergencies has always included a significant communication component in the form of warnings, risk messages, evacuation notifications, messages regarding *self efficacy*, information regarding symptoms, and medical treatment, among many others. Different kinds of crises, however, manifest different forms of threat and create different communication exigencies. Some kinds of natural disasters, such as hurricanes, usually prompt recommendations that residents stockpile water and food or that they evacuate. Effective response to hurricanes also requires careful and effective public notification of evacuations. Hazardous chemical spills often result in recommendations that residents close windows, shut off all outside ventilation, and shelter in place. Tornado warnings usually include recommendations that residents seek shelter immediately in secured locations. Floods are usually accompanied by recommendations that residents drink bottled or boiled water to avoid water-borne pathogens (Sellnow, Seeger, & Ulmer, 2002).

Public health emergencies also concern the outbreak of particular diseases or the identification of specific risks from environmental or lifestyle factors. Food-borne pathogens, such as ground beef contaminated with e.coli 0157:H7, for example, create a risk to the general public health. In such cases, public health epidemiologists track the sources of the contamination, undertake actions to stop the contamination, issue warnings and recalls, and provide the public with information about symptoms, treatments, and ways to avoid exposure (Fischhoff & Downs, 2001; Ulmer, 2001). In the outbreak of West Nile virus, as described in subsequent chapters, public health agencies have broadly disseminated information about the nature and frequency of the disease, the level of risk, and the ways to avoid exposure. Typically, these activities rely on the mass media and general public education campaigns to disseminate information and educate the public. Severe flu seasons may prompt officials to publicly offer advice on hand washing, hand sanitizers, and vaccinations. During the 2003 flu season, officials were concerned about the emergence of atypical flu and mounted a campaign calling for respiratory etiquette—covering sneezes and coughs—so as not to spread the virus. In the 2004 season, officials encountered an unanticipated shortage of vaccine and were forced to offer public explanations, describe alternative methods of protection, and issue guidelines as to whom should receive the limited supply of vaccine.

The specter of bioterrorism raised by the 9/11 attacks and the anthrax outbreak, however, raised the possibility of a new threat to the public health—widespread harm by an intentional act. Some scenarios involving agents such as botulism, plague, anthrax, smallpox, VX, mustard and Sarin gas, and radiological materials could cause massive harm to the public health (Center for Disease Control and Prevention, 2003; National Council on Radiation Protection & Measurement, 2001). Beyond the physical harm, these scenarios pose a severe threat to psychological security and the associated socio-economic stability. One of the principle ways by which this psychological harm and social disruption can be reduced is through effective communication.

Public health, then, faces a wide array of communication exigencies associated with natural disasters, emerging infectious diseases, food-borne illnesses, and bioterrorism. These communication exigencies exist at all stages of a crisis. Public messages of preparation and risk, for example, should be offered during precrisis. Communication structures, procedures, and networks should also be established as part of a preevent crisis plan. During a crisis, a variety of event-specific communication requirements arise related to reducing the uncertainty and associated anxiety and the mitigation of harm. These range widely and include recalls, evacuation notices, warnings, recommendations, and event-specific information such as disease symptoms, additional risks, road closures, and shelter information, among many others. After an event, communication usually focuses on cause and adequacy of response.

Given these complex communication responsibilities, public health officials and professional health communicators have begun to adopt new communication approaches. Many of these approaches emerged after the anthrax attack and seek to merge risk communication strategies with crisis communication practices into a form described by the Centers for Disease Control as Crisis and Emergency Risk Communication (CERC).

Risk Communication and Crisis Communication

Traditionally, a number of important distinctions have characterized the crisis communication functions of public health from more routine risk communication practices. These distinctions relate in part to the differences between risk communication and crisis communication as fields of study and areas of practice.

Risk communication is used as the basis for many public health campaigns, including those designed to promote healthy lifestyles and alert the public to environmental, lifestyle, or disease-related risks. This includes public service campaigns about smoking, safe sex, and drinking and driving; exercises, brochures, posters, and health fairs promoting disease scanning; and informational campaigns about specific risk-reduction activities, such as flu shots and infant vaccinations. The Department of Homeland Security (DHS) has adopted risk communication strategies in its Ready.Gov campaign. The National Research Council (1989) describes risk communication as "an interactive process of exchange of information and opinion among individuals, groups, and institutions" (p. 2). Risk communication, then, is the intentional effort to inform the public about risks and persuade individuals to modify their behavior to reduce risk.

Risk communication as practiced in public health often uses fear appeals as persuasive devices. Messages seek to induce behavioral change by presenting a threat and describing a behavior or behavioral change that may alleviate the threat. *Efficacy* is the effectiveness or feasibility of the behavioral change in alleviating the threat, whereas *self efficacy* refers to the belief that the recommendation can actually be carried out (Egbert & Parrott, 2001; Witte, Meyer, & Martel, 2000). Risk communication is also grounded in the public's right to know about hazards and risks that might have an impact on their wellbeing. Thus, cigarettes carry health warnings about the relationship of smoking, and companies that release carcinogens into the environment are obligated to inform the community. The availability of information allows the public to make informed choices regarding risk. In this way, risk communication facilitates decision making and risk sharing.

Sandman (2002) described the ways in which the public perceives risk based on both a technical assessment (hazard) and a culturally defined risk (outrage). Risks such as SARS and anthrax, for example, are new and unfamiliar to the general public and, therefore, are seen as unnatural and exotic, creating high levels of uncertainty. In most cases, they involve organisms that cannot be seen and diseases and symptoms not evident before in the general population. This general lack of familiarity is likely to enhance the outrage and the risk perceived by the general public (Sandman, 2002). In the case of an intentional bioterrorism attack, the perceived outrage surrounding the event will be high, further enhancing the perceived risk. Bioterrorism is also likely to create very high levels of uncertainty. As with other forms of risk communication, these emergent threats create challenges for the medical and public health community to communicate in accurate, credible, timely, and reassuring ways. Sandman's (2002) approach accommodates what is sometimes seen by public health officials as irrational or unwarranted fears. According to this model, any effort to understand the public's view of risk must include these perceptions of hazard. Sandman's work has framed many of the efforts to communicate about various public health risks.

Other basic principles of effective communication have also informed the practice of risk communication. Audiences tend to simplify messages. Credibility is important to the believability and effectiveness of messages. Risk messages should include some self efficacy—actions that can be taken to reduce the risk (Egbert & Parrott, 2001). Messages are more effective when strategically matched to audience needs, values, background, culture, and experience (Murray-Johnson, Witte, Liu, & Hubbel, 2001). Risk messages should be clear and simple, appeal to both reason and emotion, and offer solutions to problems (Friemuth et al., 2001). Much of risk communication as practiced in public health incorporates these features in public messages, usually carried in the mainstream media, as general persuasive campaigns. They seek to inform the public and change behavior in ways that protect and improve public health and safety.

In contrast to risk communication, crisis communication is more typically associated with public relations activity following an organizational crisis (see Coombs, 1999; Seeger et al., 1998). Organizational crises, such as plant explosion, employee violence, toxic spills, transportation accidents, defective products, or even management misconduct, usually generate widespread public and media interest and often criticism. Public relations provides the communication skill to strategically defend and explain the organization's position in the face of this criticism, the associated threat, and uncertainty. During crises, public relations practitioners provide accounts of what went wrong, why, and what is being done in response. This form of postcrisis public relations as a form of defensive communication has sometimes added to a cynical view of public relations. As the role of public relations has expanded and as crises have become more common, so too has the notion of crisis communication. One fundamental change involves a cardinal tenet among public relations practitioners that an honest, candid, prompt, accurate, and complete response to a crisis is always preferred (Small, 1991).

Crisis communication, then, involves the sending and receiving of messages "to prevent or lessen the negative outcomes of a crisis and thereby protect the organization, stakeholders, and/or industry from damage" (Coombs, 1999, p. 4). As such, it is part of a larger crisis management function (Seeger et al., 1998). These communication processes are designed to reduce and contain harm, provide specific information to stakeholders, initiate and enhance recovery, manage image and perceptions of blame and responsibility, repair legitimacy, generate support and assistance, explain and justify actions, apologize, and promote healing, learning, and change (Seeger, Sellnow, & Ulmer, 2003). Crisis communication seeks to explain the specific event, identify likely consequences and outcomes, and provide specific harm-reducing information to affected communities in an honest, candid, prompt, accurate, and complete manner.

Despite their differing origins, risk and crisis communication obviously have much in common and overlap in a variety ways. In fact, crisis communication can be understood as one limited form of risk communication (Lundgren, 1994). Both forms involve the production of public messages designed to influence public perceptions and behavior. In both cases, the messages are largely mediated through mass communication channels, although they also have public, organization, and group communication dimensions. Risk and crisis communication rely on credibility as a fundamental persuasive attribute, although manifest in different ways. Both share an essential purpose of seeking to limit, contain, mitigate, and reduce public harm. Beyond these basic commonalities, risk and crisis communication diverge in important ways. These differences are outlined in Table 1.1.

The basic goals of risk and crisis communication, for example, are different. Risk messages concern the probabilities of some harm and associated methods for reducing the probability of the harm. Risk messages are grounded in both current scientific, technical and empirical understanding of a specific risk factor, as well as cultural or social beliefs. Risk messages seek to translate or operationalize technical understanding of risk into behaviors through persuasive and informative messages while addressing public concern or fear. This often requires addressing the cultural or social factors related to a risk. Messages about crisis, in contrast, typically concern both what is known and what is unknown about a specific event. Often such messages are phrased specifically as "What we do not yet know about the situation" and are tempered with discussions about what is being done to collect additional information or mitigate against additional harm. Crisis messages are often more focused on informing than persuading. Risk communication has expanded as understandings of risk, particularly technical and scientific, have expanded and with increased public pressure for more information about risks. This form of communication has become common and almost routine in public service messages, brochures, and health education campaigns. Crisis communication, in contrast, has remained largely event-specific, although precrisis planning has encouraged emergency managers to move beyond the confines of a specific event. Crisis communication is essentially nonroutine and much more time- and event-bound to the specific conditions of a particular crisis. Risk communication is largely sender

TABLE 1.1.
Distinguishing Features of Crisis and Risk Communication

RISK COMMUNICATION	CRISIS COMMUNICATION
Messages regarding known probabilities of negative consequences and how they may be reduced; addressing technical understandings (hazards) and cultural beliefs (outrage)	Messages regarding current state or conditions regarding a specific event; magnitude, immediacy duration and control/remediation; cause, blame, consequences
Principally persuasive (i.e., advertising and public education campaigns)	Principally informative (i.e., news disseminated through media or broadcast through warning system)
Frequent/routine	Infrequent-nonroutine
Sender-message-centered	Receiver-situation-centered
Based on what is currently known (i.e., scientific projections)	Based on what is known and what is not known
Long term (precrisis) Message preparation (i.e., campaign)	Short term (crisis) Less preparation (i.e., responsive)
Technical expert, scientist	Authority figures/emergency manager, technical experts
Personal scope	Personal, community, or regional scope
Mediated; commercials, ads, brochures, pamphlets, Controlled and structured	Mediated; press conferences, press releases, speeches, Web sites Spontaneous and reactive

and message-centered, in the sense that time is available to carefully craft a message, identify and test appeals, and disseminate the message through targeted media channels. Crisis communication is more sender- and event-oriented, in that it seeks to respond to immediate public needs for information in what is often a much more spontaneous and less controlled manner. Because they are more spontaneous, crisis messages are typically less polished and more often take the form of news conferences or announcements covered in print and broadcast news outlets. Crisis messages also take advantage of whatever channel of communication is available at the time of the event. Radio, for example, given its flexibility and wide availability, is a particularly important medium for crisis communication. A significant proportion of the public learned about the 9/11 attacks by radio (Seeger,

Vennette, Ulmer, & Sellnow 2002). The DHS encourages individuals to have battery-operated radios as part of their personal preparedness kits.

Public health has entered a new environment where communication about both routine and well-understood threats to public health, as well as the nonroutine and poorly understood, is imperative. Among these new imperatives is the need to respond to specific crisis events within restricted time frames and address more general and often poorly understood risks. Increasingly, public health is framed as part of the first responder community, called on to take immediate action in the case of a disaster or crisis that might affect the public health. Many state and local health departments have responded to these new requirements with divisions of public health preparedness. Other public health agencies have added robust crisis communication response capabilities and are seeking to learn how to manage these events successfully.

Crisis and Emergency Risk Communication

One way the public health community has responded to these new challenges is by developing a model of communication that integrates both crisis and risk communication. Known as Crisis and Emergency Risk Communication (CERC), this approach was developed by the CDC following the anthrax episode (Reynolds, Hunter-Galdo, & Sokler, 2002). As subsequent chapters illustrate, many of the CDC's established communication practices and strategies proved inadequate during the anthrax attack. As a consequence, the CDC set out to develop new, more comprehensive approaches.

According to the CDC (2003), CERC is a method designed

> to provide information that allows an individual, stakeholder, or entire community to make the best possible decisions during a crisis emergency about their well-being and communicate those decisions, within nearly impossible time constraints, and ultimately to accept the imperfect nature of choices as the situation evolves. (p. ii)

CERC merges the strategies of risk and crisis communication by specifying a broad set of communication activities and matching these to the various stages of a crisis or disaster. The CERC model is grounded in a five-stage model of crisis, including precrisis, initial event, maintenance, resolution, and evaluation. Each stage, according to the CDC, requires specific communication activities (See Table 1.2). During precrisis, for example, public health communicators should undertake the kinds of activities most often described as risk communication. These include educating the public about risks and appropriate responses to specific adverse events, and seeking to change behaviors so that risks are reduced. In the crisis stage, a variety of communication activities such as information about the nature of the threat, responses of officials, and recommendations for individual response may be required. As the crisis moves into the resolution stage, health communicators face a different set of exigencies. At this point, the public seeks more detailed insight into the nature of the event. During the resolution stage, questions of cause, blame, and responsibility typically emerge. These questions, then, require officials to respond with a broad-based, honest, and open discussion of these issues. Finally, the evaluation stage is an opportunity to create and communicate the lessons from the crisis. Evaluation creates the opportunity for health communicators to prepare for the next infectious disease or food-borne illness outbreak. These activities are presented in Table 1.2.

TABLE 1.2.
Stages of Crisis and Emergency Risk Communication

I. **Precrisis (Risk Messages, Warnings, Preparations)**
Communication and education campaigns targeted to both the public and the response community to facilitate:

- Monitoring and recognition of emerging risks
- General public understanding of risk
- Public preparation for the possibility of an adverse event
- Changes in behavior to reduce the likelihood of harm (self-efficacy)
- Specific warnings messages regarding some eminent threat, such as evacuation notices, take shelter warnings, product recalls, etc.
- Alliances and cooperation with agencies, organizations, and groups
- Development of consensual recommendations by experts and first responders
- Message development and testing for subsequent stages

II. **Initial Event (Uncertainty Reduction, Self Efficacy, Reassurance)**
Rapid communication to the general public and to affected groups seeking to establish:

- Empathy, reassurance, and reduction in the public's emotional turmoil
- Designated crisis/agency spokespersons and formal channels and methods of communication
- General and broad-based understanding of the crisis circumstances, nature of the threat, consequences, and anticipated outcomes based on available information
- Reduction of crisis-related uncertainty
- Specific understanding of emergency management and medical community responses

- Understanding of self-efficacy and personal response activities (how/where to get more information, check on neighbors, avoid contaminated water, etc.)

III. **Maintenance (Ongoing Uncertainty Reduction, Self Efficacy, Reassurance)**
Communication to the general public and affected groups seeking to facilitate:

- More accurate public understandings of ongoing risks
- Understanding of background factors and issues
- Broad-based support and cooperation with response-and-recovery efforts
- Feedback from affected publics and correction of any misunderstandings/rumors
- Ongoing explanation and reiteration of self-efficacy and personal response activities (how/where to get more information) begun in Stage II
- Informed decision making by the public based on understanding of risks/benefits

IV. **Resolution (Updates Regarding Resolution, Discussions About Cause and New Risks/New Understandings of Risk)**
Public communication and campaigns directed toward the general public and affected groups seeking to:

- Inform and persuade about ongoing clean-up, remediation, recovery, and rebuilding efforts
- Facilitate broad-based, honest, and open discussion and resolution of issues regarding cause, blame, responsibility, and adequacy of response
- Improve/create public understanding of new risks and new understandings of risk, as well as new risk-avoidance behaviors and response procedures
- Promote the activities and capabilities of agencies and organizations to reinforce positive corporate identity and image

V. **Evaluation (Discussions of Adequacy of Response, Consensus About Lessons and New Understandings of Risks)**
Communication directed toward agencies and the response community to:

- Evaluate and assess responses, including communication effectiveness
- Document, formalize, and communicate lessons learned
- Determine specific actions to improve crisis communication and crisis response capability
- Create linkages to pre-crisis activities (Stage I)

CERC is a comprehensive and integrated approach to the communication responsibilities of public health. It seeks to represent the dynamic nature of crises, the high uncertainty and threat associated with these events, and the need to communicate within these conditions. The CDC has offered CERC training to health professionals and local officials in every U.S. state and territory. Over 100,000 individuals have been trained in CERC through Web-based and CD-Rom delivery. Another 11,000 have received training through satellite systems, and 4,500 have received the training in classroom settings. Clearly, based on reach alone, this integrated approach to communication has had a significant impact. Although the CERC framework is certainly not a panacea, and as an introductory approach does oversimplify many complex crisis-related phenomena, it does represent a comprehensive state-of-the-art methodology matched to the dynamic and threatening circumstances of a crisis. It is practitioner- and application-oriented, as opposed to creating a broad conceptual and theoretical grounding in communication and crisis. CERC does incorporate the many important communication lessons the CDC learned from managing complex events such as the anthrax attacks.

Learning From Crisis

One of most overwhelming challenges is for an organization such as the CDC to see past the immediate uncertainty and harm of a crisis to the larger lessons of preparation and response. Learning is one of the fundamental ways by which organizations, including health organizations, adjust to and accommodate the unexpected disruption of a crisis event. Learning is necessary for several reasons.

First, learning facilitates the adjustment of risk recognition and refines the collective understanding of a risk. Epidemiological processes, for example, seek to identify and trace disease outbreaks so they can be more effectively contained and controlled. These processes often result in the identification of new routes of infection. Food-borne illness, as subsequent chapters illustrate, often occurs as a result of poor personal hygiene, food being washed in dirty water, cross-contamination, or poor storage. Learning requires identification of these sources, often resulting in seeing something as a risk factor for the first time.

Second, norms and avoidance systems may be adapted so that the probability of a repeat event is reduced. Once risk factors such as routes of infection have been identified, actions can be taken to ensure that future infections via these routes are reduced or eliminated. This may involve new procedures, policies, structures, or operations. Several cases of SARS among health care workers, for example, were traced to exposure in ambulances or emergency rooms. This resulted in new procedures for handling SARS patients.

Third, learning may result in new resources or capacities. If a crisis is not managed successfully, agencies often face severe criticism, particularly in the postcrisis stage. This criticism may take the form of a detailed postmortem and investigation followed by changes in personnel, restructuring of operations, and commitment to operate differently. In some cases, organizations associated with a crisis, or failed management of a crisis may be forced to overhaul operations, reorganize, change personnel, or give up authority.

Crisis-induced lessons can be quite powerful forces in organizational or agency change. The events of 9/11, for example, created the largest ever reorganization of the federal bureaucracy, leading to the creation of the new DHS. Such a comprehensive change could not have been possible without the impetus of 9/11. Hurricane Katrina created a new set of lessons for DHS. Among other things, Katrina and the 2005 hurricane season demonstrated the fallacy of many optimistic planning assumptions. In addition, the CDC instituted a number of changes following the anthrax episode. Vanderford (2003) noted that the failure of the CDC had more to do with inadequate application of principles of communication than with inadequate knowledge or skill. Similarly, Robinson and Newstetter observe that the anthrax episode fundamentally altered the CDC's processes of producing and disseminating information. Many of the communication lessons learned by the public health community as they plan for and respond to crises are outlined in the following chapters.

Overview of Chapters

Part II of this book focuses on bioterrorism through the case of the anthrax attacks. Sandra Quinn and her colleagues (chap. 2) offer a detailed analysis of the anthrax episode based on interviews with postal workers and public health professionals. Julie Novak and Scott Barrett (chap. 3) conducted a content analysis of media coverage regarding anthrax and concluded that too many spokespersons created a confused and inconsistent message.

Part III concerns communication issues in outbreaks of food-borne illness. This form of crisis has traditionally been one of the most harmful to public health. The advent of modern food production and processing technology has reduced many of the threats, but introduced new ones through complexity, mutually interdependent systems, and global food production. The problems of producing food safely are explored by Steve Venette and his colleagues (chap. 4) in an analysis of a turkey processing plant. Denise Gorsline and Sheri Erickson (chap. 5) examine the case of food-born illness traced to green onions used at a Chi-Chi's Mexican restaurant. The case illustrates the vulnerabilities built into a global food production and distribution network.

Part IV includes chapters that explore infectious disease outbreaks. Robert Ulmer, Robert Alvey and Jennifer Kordsmeier (chap. 6) examine public health campaigns designed to address West Nile virus. Two chapters focus on the important case of sudden acute respiratory syndrome (SARS). Min Liu, J. J. McIntyre, and Timothy Sellnow (chap. 7) examine the problems of reporting and tracing disease outbreaks in countries where press freedom is restricted. Robert L. Heath, Feng Li, Shannon A. Bowen, and Jaesub Lee (chap. 9) argue that, because of systemic, rhetorical, and ethical flaws, the SARS narrative was incomplete, slow to develop, and framed defensively. These flaws contributed to the inadequate and slow response. Keith Hearit (chap. 9) offers an insightful analysis of the metaphors associated with public health communication through an examination of an encephalitis outbreak. Finally, Avinash Thombre (chap. 10) examines the intersection of crisis and risk communication in his analysis of an HIV/AIDS campaign in India. Thombre suggests that disease stigmatization is one factor allowing a risk to incubate into a crisis.

Part V explores the ways in which the public health and emergency management communities have responded to these contingencies through crisis preparation and response. Deborah Ballard-Reisch and her colleagues (chap. 11) conducted an assessment of crisis preparedness in one state using the CERC model. Their analysis includes support for the CERC model as well as a refinement and extension of this framework. Timothy Coombs (chap. 12) explores the DHS' public preparedness campaign using the Parallel Process Model. He concludes that DHS was successful in creating, public fear as a motivating state, but was unable to instill any sense of self efficacy. Scott R. Magee, J. Gregory Payne, and Scott Ratzan (chap. 13) address the fundamental questions of the ethical issues in crisis communication. Crisis and disasters often place ethical issues and values at the forefront of responses. Further, a crisis response judged as ethical is often more effective than those seen as self-serving or deceptive. Finally, Mike Lewis (chap. 14) explores the problems of media management in public health reporting. His analysis includes an analysis of several of the CERC strategies for media relations.

CONCLUSION

We began this introductory chapter by enumerating the extraordinary range of risks and threats that face society today. These threats are dynamic, global, and becoming increasingly prominent. Successful preparation for and management of these threats represents an immediate imperative for public health and the larger emergency management community. A significant component of successful management is a more sophisticated, dynamic, and comprehensive approach to communication. This includes ongoing communication with the public, those diverse stakeholder groups affected by the crisis, the media, government at all levels, response agencies, and communities. The approaches outlined here begin to address these crisis and risk communication imperatives. These approaches are an acknowledgment that risks are more ubiquitous and that emergencies and crises threatening the public health and well-being are likely to be increasingly common. Moreover, as illustrated in the following chapters, any successful effort to reduce the threats and respond to crises must have a central focus on communication.

REFERENCES

Centers for Disease Control and Prevention. (2003). Retrieved April 20, 2003, from http://www.cdc. gov

Coombs, W. T. (1995). The development of guidelines for the selection of the "appropriate" crisis response strategies. *Management Communication Quarterly, 4,* 447–476.

Covello, V. T. (1992). Risk communication: An emerging area of health communication research. In S. A. Deetz (Ed.), *Communication Yearbook 15* (pp. 359–373). Newbury Park, CA: Sage.

Department of Homeland Security. (2003). Retrieved April 13, 2003, from http://www.ready.gov

Egbert, N., & Parrott, R. (2001). Self-efficacy and rural women's performance of breast and cervical cancer detection practices. *Journal of Health Communication, 6,* 219–233.

Fischhoff, B., & Downs, J. S. (2001). Communicating foodborne disease risk. *Emerging Infectious Diseases, 3,* 4. Available from http://www.cdc.gov/ncidod/eid/vol3no4/fischof.htm

Heath, R. L. (1994). Environmental risk communication: Cases and practices along the Texas Gulf coast. In B. R. Burelson (Ed.), *Communication Yearbook 18* (pp. 225–277). Newbury Park, CA: Sage.

Lundgren, R. E. (1994). *Risk communication: A handbook for communicating environmental, safety and health risks.* Columbus, OH: Battelle.

Murray-Johnson, L., Witte, K., Liu, W., & Hubbel, A. P. (2001). Addressing cultural orientations in fear appeals: Promoting AIDS-protective behaviors among Mexican immigrant and African American adolescents and American and Taiwanese college students. *Journal of Health Communication, 6,* 335–358.

National Council on Radiation Protection & Measurement. (2001). *Management of terrorist events involving radiological materials.* Bethesda, MD: Author.

National Research Council. (1989). *Improving risk communication.* Washington, DC: National Academy Press.

Reynolds, B., Hunter-Galdo, J., & Sokler, L. (2002). *Crisis and emergency risk communication.* Atlanta, GA: Centers for Disease Control and Prevention.

Sandman, P. M. (1993). *Responding to community outrage: Strategies for effective risk communication.* Fairfax, VA: American Industrial Hygiene Association.

Sandman, P. M. (2002). Smallpox vaccination: Some risk communication linchpins. Retrieved May 4, 2003, from http://www.psandman.com/col/SARS-1.htm

Sandman, P. (2002). "Fear is spreading faster than SARS"—And so it should. Retrieved May 4, 2003, from http://www.psandman.com/col/SARS-1.htm

Seeger, M. W., Sellnow, T. L., & Ulmer, R. R. (1998). Communication, organization and crisis. In M. E. Roloff (Ed.), *Communication Yearbook 21*. Thousand Oaks, CA: Sage.

Seeger, M. W., Sellnow, T. L., & Ulmer, R. R. (2001). Public relations and crisis communication: Organizing and chaos. In R. L. Heath (Ed.), *Public relations handbook* (pp. 155–166). Thousand Oaks, CA: Sage.

Seeger, M. W., Vennette, S., Ulmer, R. R., & Sellnow, T. L. (2002). Media use, information seeking, and reported needs in post crisis contexts. In B. S. Greenberg (Ed.), *Communication and terrorism: Public and media responses to 9/11* (pp. 53–63). Cresskill, NJ: Hampton Press.

Sellnow, T. L., & Ulmer, R. R. (1995). Ambiguous argument as advocacy in organizational crisis communication. *Argumentation and Advocacy, 31,* 138–150.

Sellnow, T. L., Seeger, M., & Ulmer, R. R. (2002). Chaos theory, informational needs and the North Dakota floods. *Journal of Applied Communication Research, 30,* 269–292.

Small, W. (1991). Exxon Valdez: How to spend billions and still get a black eye. *Public Relations Review, 17*(1), 9–26.

Ulmer, R. R., & Sellnow, T. L. (2000). Consistent questions of ambiguity in organizational crisis communication: Jack in the Box as a case study. *Journal of Business Ethics, 25*(2), 143–155.

Vanderford, M. L. (2003). Communication lessons learned in the Emergency Operations Center during the CDC's anthrax response: A commentary. *Journal of Health Communication, 7,* 11–12.

Witte, K., Meyer, G., & Martel, D. (2000). *Effective health risk messages.* Thousand Oaks, CA: Sage.

PART II

Bioterrorism

2

Lessons From the 2001 Anthrax Attack

A Conceptual Model for Crisis and Emergency Risk Communication

Sandra Crouse Quinn

Tammy Thomas

Carol McAllister

This chapter reports on the results of interviews with postal workers and public health professionals involved in the anthrax attacks to explore theories relevant to a bioterrorist crisis and propose a new model. Although some of the theories suggested in the literature are, in fact, relevant in a crisis, other critical factors may be pertinent to any given bioterrorist event. Therefore, a conceptual model may offer a more flexible approach in the response, thereby permitting each response to reflect appropriately the needs of those affected.

Trying to take issues . . . complicated public health science and how do you translate that into everyday language that people can understand so they can make informed decisions and judgments about what actions they needed to take.

–Reflections of a public health professional

Communication—a process by which information is exchanged between individuals through a common system of symbols, signs, or behavior—is difficult and challenging even in the best of circumstances (Merriam-Webster Online, 2005). However, it becomes profoundly problematic during a crisis. This chapter addresses communication during the 2001 anthrax attack and proposes a conceptual model encompassing existing theories and results from a qualitative study of postal workers and public health professionals.

The use of anthrax as a bioterrorist agent brought to light the unique communication challenges that emerge from this type of emergency. Although the science of risk communication has been evolving, risk communication theories have not specifically addressed the demands of a terrorist event. Based on the events of 2001, the Centers for Disease Control and Prevention (CDC) recognized the need to develop crisis and emergency risk communication (CERC), which they defined as,

> an effort by experts to provide information to allow an individual, stakeholder, or an entire community to make the best possible decisions about their well-being within nearly impossible time constraints and help people ultimately to accept the imperfect nature of choices during the crisis. (Centers of Disease Control, 2002a, p. 6)

CERC builds on the existing literature in the field of health and risk communication and synthesizes those with crisis communication (Reynolds & Seeger, 2005). Risk communication has historically focused on community environmental health concerns, such as the placement of hazardous industries, potential or real environmental emergencies, and potential exposures to toxins. More recently, risk communication has expanded to address issues related to personal health decisions. The literature on risk communication theory focuses on risk perception, heuristic processing of risk information, mental models and other theories (Covello & Sandman, 2002; Fischoff, Bostrom, & Quadrel, 1993; National Research Council, 1989; Reynolds & Seeger, 2005). *Risk communication* is defined as,

> an interactive process of exchange of information and opinion among individuals, groups, and institutions. It involves multiple messages about the nature of risk and other messages, not strictly about risk, that express concerns, opinions, or reactions to risk messages or to legal and institutional arrangements for risk management. (National Research Council, 1989, p. 21)

Crisis communication can be defined as the communication of messages "to prevent or lessen the negative outcomes of a crisis and thereby protect the organization, stakeholders and/or industry from damage" (Coombs, 1999, p. 4). Although its origins are in public relations, crisis communication has expanded its scope to address natural disasters.

The CERC approach is distinctive from health, risk and crisis communication in four key ways: (a) decisions must occur in a compressed timeframe; (b) the decision may be irreversible; (c) the outcome uncertain; and (d) the information necessary for that decision may be incomplete or uncertain (Centers for Disease Control, 2002). CERC recognizes a process that includes preevent, emergence of the event, crisis period, and follow-up and evaluation (Reynolds & Seeger, 2005). Although CERC offers a model for planning, organizing and delivering communication, it does not explicitly use a theoretical foundation for that communication. Therefore, the development of a conceptual model can provide further guidance for communication during a bioterrorist event.

The growing literature examining the challenges of communication during a terrorist event argues that successful communication during a crisis is complex and will require addressing multiple factors (Covello et al., 2002; Fischoff, 2002; Sandman, 2003; Sorenson, 2004; Wray et al., 2004). Consequently, the question of which theories may be appropriate to inform CERC arises. Some researchers argue that communication during a crisis must draw on multiple theories. Covello and colleagues (2002) assert that the mental noise model, risk perception, the trust determination model, and negative dominance theories are all pertinent to communication during a bioterrorist event. Wray and colleagues (2004) examine communication from a socioecological perspective, ultimately recommending that different theories are necessary for communication to the broader population and for changing individual behavior in a crisis. They suggest the Social Amplification of Risk model as the foundation for the population-level communication, with McGuire's persuasion theory, Elaboration Likelihood Theory, risk perceptions, and behavioral theories as useful for ensuring that individuals adopt the necessary behaviors for the crisis.

We agree that no single existing theory is adequate as the foundation of CERC. Thus far, articles addressing theory have not been informed by those who have actually experienced a bioterrorist event. In this chapter, we build on the results of interviews with postal workers and public health professionals involved in the anthrax attacks to explore the theories relevant to a bioterrorist crisis and propose a new model. Although some of the theories suggested in the literature are, in fact, relevant in a crisis, other critical factors may be pertinent to any given bioterrorist event. Therefore, a conceptual model may offer a more flexible approach in the response, thereby permitting each response to reflect appropriately the needs of those affected.

The value of a conceptual model is that it blends theoretical constructs and actual experience to offer new guidance for professionals charged with communicating during a crisis. Earp and Ennett (1991) define a *conceptual model* "as a diagram of proposed causal linkages among a set of concepts believed to be related to a particular public health problem" (p. x). Therefore, where a theory focuses on explanations of broader classes of phenomena, a model is concerned with specific behaviors that occur within a specific context. Earp and Ennett described the advantages of a conceptual model: (a) it can be informed by more than one theory, (b) it can be inclusive of different levels of intervention from a micro- to macroapproach, and (c) it allows inclusion of processes or characteristics from empirical findings or experience. In this case, we examine the anthrax attack of 2001 to weave a tapestry of theories and empirical experiences into a synergistic model.

THE ANTHRAX ATTACK OF 2001

In October 2001, the United States was reeling from the attack on the World Trade Center and the Pentagon and the crash of Flight 93 in Pennsylvania. Then, in the midst of this national trauma, anthrax sent through the U.S. Postal Service (USPS) caused 22 cases of the disease, including five deaths. Nine cases were postal employees, including six diagnosed with inhalational anthrax (Jernigan et al., 2002). The Hart Senate Office Building, where anthrax was released from a letter to Senator Tom Daschle, closed October 17, 2001, and the following day, the Trenton Postal Processing and Distribution Center (PDC), where two postal workers were diagnosed with inhalational anthrax, closed (Greene et al., 2002; Tan et al., 2002). Between October 19 and 21, 2001, four workers at the Brentwood PDC in the District of Columbia were hospitalized with inhalational anthrax; two African American workers died. The building was closed on October 21 (Centers for Disease Control and Prevention, 2001). Anthrax was found in the Morgan Central Postal Facility in New York City, but that facility remained open. Contamination at the Morgan facility was considered isolated so decontamination occurred while operations continued elsewhere in the facility (Government Accountability Office, 2004). Anthrax contamination was ultimately found in twenty-one postal facilities (Government Accountability Office, 2004). Public health professionals faced significant challenges in communicating to postal workers, health care providers, elected officials and the broader public about this threat because the events created an endlessly changing picture.

LISTENING TO THE VOICES OF THOSE AFFECTED: A CDC STUDY

In 2002, the CDC, through its cooperative agreement with the Association of Schools of Public Health, funded a qualitative study of postal workers and public health professionals in Washington, DC, New Jersey, and New York City. That study included: nine interviews with key informants among postal workers in DC, NJ, and NY; 20 additional interviews with individual postal workers in DC and NJ; six focus groups with postal workers across the three sites; and 16 interviews with public health professionals (8 were from local or state health departments and 8 were CDC staff). In addition, we attended several meetings of Brentwood Exposed, an activist group of postal workers formed in response to the anthrax crisis in DC. All interviews and focus groups were taped and transcribed verbatim. Two researchers coded the transcripts identifying key issues and themes, while a third assisted with the interpretation of findings. Although this qualitative study does not allow for generalization to all postal workers, the finding of consistent themes across the three sites, as well as between the interviews and focus groups, provides confidence in the results.

The voices of those who lived through the anthrax attack provide the foundation for our examination of existing theories and proposal for a conceptual model. This chapter offers brief descriptions of the existing theories particularly relevant to the anthrax attack. Quotations from the respondents illustrate the salience of the theory to a conceptual model. Finally, the perspectives of the participants suggest additional factors that will, with the theoretical components, comprise a new conceptual model for CERC.

Social and Environmental Context

Social Amplification of Risk is "based on the thesis that events pertaining to hazards interact with psychological, social, institutional and cultural processes in ways that can heighten or attenuate individual and social perceptions of risk and shape risk behavior" (Renn et al., 1992, p. 139). The model recognizes that individuals exist as members of larger groups who "codetermine the dynamics and social processing of risk," and that the context in which they perceive risk or, as they describe it, "the rules of their home organization" affect their perception (Renn et al., 1992, p. 141). Although this model recognizes the importance of numerous factors in the experience of risk, it appears, despite its attempt to reconnect through its feedback loop, to be somewhat limited by its linear progression from sources to receiver.

The Social Amplification of Risk model places the receiver (audiences) at the end of the communication process. Although there is recognition that the social context in which a hazard occurs affects the experience of risk and the feedback loop by which the receiver can influence the source (public health agency or other agency), the early messages may not be well informed by knowledge of the audiences (Pidgeon, Henwood, & Maguire, 2001; Renn et al., 1992; Slovic, 2002; Wray et al., 2004). This is in contrast to Covello's (2003) assertion that best practices in CERC include "listening to people," specifically by conducting formative research prior to communication activities, and empathizing with one's audience.

In the context of the anthrax attack, the need to communicate quickly and constraints on federal agencies influenced public health professionals' ability to understand the social and environmental settings in which the attack occurred. This context set the stage for the interaction between postal workers and public health professionals. Prior to the anthrax attack, there was a long-standing hostile and dysfunctional relationship between management at the USPS and labor unions; this relationship was characterized by significant distrust and anger. Public health professionals, by necessity, partnered early on with the USPS. However, without adequate formative research to understand the context, the atmosphere of distrust and anger toward USPS created significant barriers for public health professionals. Postal workers viewed health officials as partners with USPS management: "I'm still angered over the fact that they are in cahoots with the postal service. So did we trust them? No." This shaped the reception professionals received. One professional described a common experience: "I was met with a lot of anger from the very first, and it was one of the most antagonistic situations I have ever walked into." Other public health professionals agreed with this comment; "There is so much rage at the people who manage the post office . . . when they call a meeting and they introduce you, that's it. You're absolutely one hundred percent a target." Overcoming this distrust was never fully accomplished. However, well into the crisis in late November and December, the efforts of CDC and New Jersey health educators to reach out directly to postal workers and actively partner with one strong union leader in New Jersey helped to moderate some of the anger at that site.

The physical environment in the facilities led many postal workers to suspect that they were at more risk than public health professionals initially understood. The normal operations of the facilities included processes for cleaning machinery that distributed dust and contaminants from those machines into the air. Although early on public health professionals were not fully aware of the operations that created additional risk, many postal workers recognized this risk and responded with a lack of confidence in the risk-reduction messages they were receiving. One worker described it this way; "Wash your hands. This doesn't make any sense to me. We are at great risk here. We used to blow the machines out."

Class and Race

A growing literature suggests that issues of race and class are key components in CERC (Blanchard et al., 2004; Quinn, Thomas, & McAllister, in press; Slovic, 1999; Sorenson, 2004; Stein et al., 2005). Recognizing that race, class and gender, among other factors, affect risk communication, Slovic (2002) suggests that, in addition to a more typical risk analysis, we must consider "'vulnerability analysis,' which characterizes the forms of physical, social, political, economic, cultural, and psychological harms to which individuals and modern societies are susceptible" (p. 425). He goes on to suggest that social inequalities are one of the primary causes of vulnerability. In the context of this attack, the postal workers argue that their race, class, and occupational status placed them in harm's way to a greater degree than White congressional members and staff.

Class and race played multiple roles in the anthrax crisis and affected postal workers' perceptions of fairness or equity. Even prior to the attack, the divisions between labor and management mirrored race and class differences and contributed to the tension during the crisis. Reflecting on a meeting between union and management officials, one public health professional made this observation; "It seems almost like class warfare that goes on there." Race was an issue, particularly in the Brentwood facility, as reflected in this postal worker's comment: "It doesn't help that all supervisors are White and the postal service is 97 percent African American."

The extent to which most postal workers viewed their class and race as a critical factor in the crisis was illustrated by their reaction to the closing of the Hart Senate Office Building. Uniformly, postal workers felt the USPS' failure to respond as quickly with closure of postal facilities reflected their "second-class" status compared with the largely white, professional workers on Capitol Hill. Most workers in the three sites felt this sense of second-class citizenship regardless of race, ". . . [We] felt like second hand . . . you know, compared to the people at the Capitol. We felt like second-hand laundry or second-hand clothing or however you want to describe it, it felt that we were second to them."

Other postal workers believed that race was clearly a factor, as reflected in this comment; "We felt . . . that the people at the Capitol were important but the Black people over there at Brentwood, which is 99.9% Black, didn't matter." "We have the majority Black and Hispanics. If there were White people, they would have closed it down," workers in New York echoed. These findings are not surprising given Slovic's (1999) commentary on risk and trust: "Whoever controls the definition of risk controls the rational solution to the problem at hand. If you define it [risk] another way, perhaps incorporating qualitative characteristics and other contextual factors, you will likely get a different ordering of your action solutions. Defining risk is thus an exercise in power" (p. 699).

Social Network Contagion

In the Social Network Contagion model, Scherer and Cho (2002) argue that individuals in frequent interaction with others constitute networks that have their own cultural norms, expectations, knowledge, and behavioral support. They suggest that, within the network, members will share and perhaps create perceptions of risk. In the anthrax attack, the postal workers undoubtedly communicated information through their networks with both positive and negative results. In some cases, the communication helped postal workers increase their knowledge of their risk and appropriate precautions. Workers across sites described a process whereby "you called somebody, they call somebody, they called somebody" as a means to get information to colleagues. In one site, a union leader developed a Web site and listserve to distribute communications to his members. He reports using health departments, media, and CDC sites to help ensure that he had credible information. Consequently, postal workers identified him as a trustworthy source.

Unfortunately, the close network also spread inaccurate or misleading information. One worker commented, "Some of it was word of mouth from other people to the point where who do you believe and who don't you believe." One public health professional commented on the interaction between the social context of the attack with the potential ability to negatively affect communication,

> You have this issue where you are going to have a lot of anger getting fueled by this background tension and you are going to have people walk in there and give some theories that fuel the energy. Misery loves company so you'll have some in there giving a conspiracy theory. What a harmful thing that is to do.

In reflecting on the anthrax attack, one professional offered a suggestion for working with social networks to foster productive CERC. He said, "If we could develop this kind of community vulnerability assessment, then you could create lines of communication. When information comes in, you could feed that line and then people could reach out to whom they trust . . . like their clergy, like their pharmacist. Give them the right messages." Communicating messages effectively is linked to understanding the health literacy and cultural and linguistic needs of the audience.

Health Literacy and Cultural Sensitivity

Health literacy is defined as the evolving skills and competencies needed to find, comprehend, evaluate, and use health information and concepts to make educated choices, reduce health risks, and improve quality of life (Zarcodoolas, Pleasant, & Greer, 2002). However, in the midst of the anthrax crisis, doing the typical formative research to help public health professionals better understand their audiences and effectively tailor messages was challenging.

Although many postal workers reported getting anthrax information from the media, their networks, and other sources, they did not always know how to evaluate the credibility of the information. Furthermore, the complexity of communication during the anthrax crisis left many postal workers without a basic understanding of anthrax. Consequently, many workers said: "Forget the scientific jargon, the medical stuff, just come in and let us know in easy terms what the heck is going on." One public health professional commented,

> I thought that my responsibility was to learn how to educate and not speak in jargon terms and when I did that, they got it. It's when I would go into my jargon place that they would miss what I was saying.

However, public health agencies were sometimes challenged by the need to develop materials that were suitable to the postal worker's literacy. One public health professional talked about the challenge of helping scientists within their agency understand the nature of postal workers' concerns; "We were trying to explain to them that these are the questions we get. We have to explain it in a way that's going to make sense to a lay person. They made it like it was a journal article." The conflict over the need to translate the science into terms appropriate for the postal workers reflected the different perceptions that professionals had of their audience. One professional reported,

> We did the health education things and they were going through "what does prophylaxis mean?" They had all these ridiculous words in there; they were calling it Bacillus anthracis in the health education document. When I asked them, if you are not going to use the word, Anthrax, can you at least put it in parentheses that B. anthracis is the same thing as Anthrax. The woman said it would be insulting to the postal workers if we did that.

Another professional commented,

> You need multiple ways to convey this information because it's complex. We don't even get it and it's detailed and not everyone is going to understand it if they hear it orally. Some people are going to like it written. Other people are going to like it visually.

Coupled with the complex nature of information about anthrax was the need to communicate with multiple different audiences within the postal service, including the hearing impaired, who constitute a substantive proportion of workers. One professional reported:

> One of the lessons we were badly in need of was to be considerate of people from either ethnic background. . . . We didn't produce any of our materials in languages other than English. We found out that the postal facilities employ a phenomenal number of deaf and mute people, and when management held their meetings, they never ever brought in a signer (interpreter). We made sure we had a signer that could help sign and that won a lot of credibility.

The Social Amplification of Risk model allows for inclusion of these critical components in the CERC process. This gestalt sets the stage of the next set of theories that resonate with the lived experiences of those affected by the attack.

Mental Noise

The Mental Noise model posits that, during a crisis event, there is a significant amount of stress that interferes with information processing (Covello, Peters, Wojtecki, & Hyde, 2001). Empirical evidence shows that when people are experiencing high levels of stress, especially when they perceive serious threat, they exhibit impairment in their ability to process information clearly and efficiently. Therefore, to enable the public to process health and safety information, the Mental Noise model proposes specific guidelines for message development in order to enhance the understanding of messages. Covello (Centers for Disease Control and Prevention, 2002b) suggests that, to counter mental noise in a crisis situation, communication should include a limited number of brief messages, which are ideally less than 9 seconds and 27 words for three key messages; use graphics and clearly understandable language at the sixth- to-eighth-grade readability level; and balance negative messages with solution-oriented messages.

Some public health professionals observed the effect of mental noise on the postal workers. One described it this way: "It was such a scary thing that it created this mental noise. Their minds were just racing and they were scared. I don't think they were absorbing a lot of what was being told to them. They needed to hear these things over and over again." This was confirmed by another professional in a different site who said, "People didn't retain information, considering the stress of the situation." However, another professional recognized that being attentive to the issue of mental noise and listening to the workers helped them improve the communication; "We found out we would get the same questions over and over so we would incorporate that into the presentation."

Both public health professionals and postal workers reported that people needed information reviewed repeatedly so that information could be absorbed. "Written information is better," one worker noted, "because you and I know, once you tell somebody something, they are liable to either forget it or they will change it."

According to participants, the challenge posed by mental noise was only increased by the number of different sources of communication and the mixed messages they received. One public health professional reported, "The FBI had been there, the postal union had something for them, CDC had given them a message. Everybody was giving them messages but not all of them were uniform." A postal worker confirmed this: "The news gave me one thing and then management and union gave me something else."

One participant observed the contradictory messages between the actions of public health professionals and their verbal messages early in the response:

> I felt angry and betrayed because there was a point when we would be at work and you would see people come in hazmat uniforms. You come in there in a hazmat uniform walking right by me and I am working. There is a problem there. I am doomed.

Postal workers were still trying to process and understand the initial messages when new evidence prompted a whole new set of messages. Especially for postal workers who were feeling very fearful, this process became overwhelming. One participant described this, "You were getting information that was contradicting what they already said. This got a little scary." One example of this challenge was the communication about the anthrax vaccine, as related by one participant: "They are disagreeing about shots and not giving shots and not doing this and not doing that, I don't know. It kind of scared me."

Other public health professionals confirmed that workers were challenged by the amount of new information, the lack of consistency in messages, and the extreme stress. Consistent with Covello, they suggested that messages needed to be repeated and constructed in clear, simple language. Contributing to the mental noise experienced by postal workers were factors affecting how they perceived their risk.

Risk Perception Factors

Although there is significant literature on factors that affect risk perception, little of that is based on a bioterrorist event. Covello and Sandman (2002) identify numerous factors that affect risk perception. They argue that these factors affect whether an audience perceives a greater or lesser degree of risk, and they contribute to anger, fear, worry, anxiety, hostility, and concern. These factors, often termed *outrage factors*, combine with the hazard to create one's perception of the risk. In the context of bioterrorism, Covello et al. (2001) assert: ". . . [O]utrage factors take on strong moral and emotional overtones, predisposing an individual to react emotionally, which in turn can significantly amplify levels of perceived risk" (p. 384). They also suggest that in the event of a bioterrorist attack, the full set of risk perception factors would be triggered, resulting in intensified perceived risk. Reynolds and Seeger (2005) concur, emphasizing that lack of familiarity, the exotic nature of the threat, and uncertainty are key factors that accentuate risk.

Indeed, the majority of the risk perception factors were triggered by the anthrax attack. In Table 2.1, quotations from postal workers or public health professionals are linked to specific factors that arose consistently across the three sites. This cacophony of risk perception factors contributed to the heightened sense of risk and affected the mental noise that influenced postal workers' responses.

Uncertainty, which is likely to be a factor in any terrorist event, was a major hurdle in the anthrax crisis. At some points, the continuing investigation meant that professionals did not always have an answer for the anxious postal workers. Many postal workers resonated with this quotation, "I don't know if it is a protection mechanism. For some reason, everybody is embarrassed to say in the beginning, I don't know. What's wrong with saying I don't know and then go do your research?" However, one postal worker went even further, "If you don't know, say you don't know and tell me you are just as afraid too. Don't give me the impression that everything is okay." Conversely, other postal workers wanted information quickly and responded to public health professionals' uncertainty with anger and distrust.

The changing and uncertain messages impacted the credibility of public health professionals and contributed to the ongoing stress and anger of many postal workers, affecting their trust in public health professionals.

Trust Determination Model

Within the context of a terrorist event, trust in the public health professional's communication is critical to ensuring appropriate actions to prevent harm or mitigate fears (Centers for Disease Control, 2002b; Covello et al., 2001; Shore, 2003; Sorenson, 2004; Working Group on Governance Dilemmas in Bioterrorism Response, 2004). The Trust Determination model describes four key elements: caring and empathy, competence and expertise, dedication and commitment, and honesty and openness (Covello et al., 2001). Disagreement among experts, lack of coordination among organiza-

TABLE 2.1.
Quotations on Factors Affecting Level of Perceived Risk

RISK PERCEPTION FACTORS	VOICES OF POSTAL WORKERS AND PUBLIC HEALTH PROFESSIONALS
Human origin	"This was weaponized anthrax."
	"They have not caught the person or persons who were responsible. Now to find out that this weapons-grade anthrax was only made at Ft. Detrick, a government installation. Now things are starting to go through our minds. Was it something from the government? We just don't know."
Involuntary	"They sent us on a suicide mission without our permission."
	"We were forced to work under those conditions. They didn't give us an option."
Uncertain or unknown risk	"Uncertainty, things were bouncing around so quickly."
	"Knowledge did change daily, more than daily at some point so that made the postal workers very skeptical. They were like well, what are we supposed to believe? What are you going to say to us tomorrow? How do we know that what you are telling us now is right?"
Lack of fairness or equity	"If they had closed that building when the Senate closed theirs, these people would still be here."
Dread [of risk from activities that evoke fear, terror or anxiety]	"You come in there in a hazmat uniform walking right by me and I am working. There is a problem there. I am doomed."
Lack of familiarity	"I didn't understand when I heard this thing, anthrax."
	"For a lot of us, this was the first time hearing about it. What is anthrax?"
	"A risk level, for an untrained person to assess their risk level. I mean, we are not scientists or physicists or anything like that. I considered it high because unknown risk is high."
Poorly understood risk	"They told us about the 8,000 spores. What's a spore? It's so vague that you don't know what you are dealing with."

Lack of trust in institutions associated with the risk	"I didn't trust them at all to insure our safety. I did not trust them at all."
Risks from activities in which one has a personal stake	"My family was my main concern in my mind. I have to survive to be with them."
Risk under the control of others	"Even if they left the building open, they should have at least given us prevention."
Catastrophic potential for death and injury to significant number in short time period or space	"It could kill me. That scared me, that really scared me and I felt like I shouldn't be put in that situation, me or nobody else. That's what really scared me. I didn't like that my life was in danger."
Delayed effects	"I don't know if any of it or all of it or none of it could be ascribed to heavy duty antibiotics for 100 days or exposing to anthrax or something else or some combination. . . ."
Significant media attention	"It was all over the news . . . on the radio, CNN. . . ."

Note. Risk perception factors are taken from "Risk Communication: Evolution and Revolution" by V. Covello and P. Sandman, in Solutions to an Environment in Peril, by A. Wolbarst, 2001, Baltimore, MD: Johns Hopkins University Press. © 2001 by Johns Hopkins University Press. Reprinted with permission.

tions addressing the event, unwillingness to acknowledge risks or disclose information, and perceived irresponsibility in managing risk and authorities' insensitivity to the public's need for dialogue and participation can decrease trust (Covello et al., 2001). Social trust in institutions may be particularly important in unfamiliar events; in fact, Frewer (2004) suggests that "Trust is likely to be particularly important under circumstances where people feel they have very little personal control over their personal exposure to potential hazards" (p. 393).

Covello (2003) considers demonstrations of caring and empathy as critical best practices in a crisis situation. Unfortunately, many postal workers did not perceive that public health agencies cared about them. One postal worker echoes the sentiment of many; "So if there was more communication with us from them, from all the different agencies, we would have felt better. We would have felt more like we were being looked after. That somebody gave a damn. We didn't feel that way at all." A public health professional observed this need in postal workers: "Who is going to hold my hand? Who is going to be there for me at my time of need because if you are not going to be there for me in my time of need, then I don't know that I can trust you?"

Another critical component of the Trust Determination model is competence and expertise. Postal workers perceived that public health professionals lacked the knowledge and expertise to assist them in the crisis. A common sentiment was, "Why didn't CDC have more information because they have been studying this for over 20 years?" The perceived lack of expertise was compounded when officials held meetings and "could never really answer the questions. They would always say, 'Well, we'll get back to you on that,' and they would never get back to us."

Consequently, some public health professionals were cognizant of needing to demonstrate their commitment to the postal workers. One commented, "When they asked us to prepare something for them, something in writing, we did it." Another said, "The key is, in my opinion, if you can put their minds at ease and say we are going to be back here 3:00 every single day. We'll be here at 3 so bring your questions down." According to one professional, the lack of commitment and dedication had a devastating effect on postal workers; "You are telling me you are going to do what it takes to help me. Then you run out the door or you make yourself so inaccessible that it is effectively running out the door. I don't really think I'm going to be helped by you."

The perceived lack of caring and empathy was compounded by what workers perceived as a lack of honesty and openness during the crisis; "They are underestimating citizens. If you be honest with people, people have a tendency to take risk as long as they know what they are dealing with and what can happen to them if they deal with it." One public health professional observed that multiple factors affected trust,

> The first thing that definitely affected trust right off the bat was being lied to. The second thing that affected trust was incomplete information. I would say that incomplete information and the lack of empathy were strong determinants of trust.

Mixed messages from different agencies also damaged trust. One postal worker observed, "You still don't know because you don't know who to believe. You have two companies that are reliable or credible and they are in disagreement." The coupling of these factors with the partnership with USPS contributed to significant distrust of public health professionals. One worker summed this up: "It was my employer that I didn't trust first. Then it just rolled into not trusting the public health system because they are all part of the same government." In this context, participation of the postal workers in the communication process was limited.

Participation

In the risk communication literature, there is a significant focus on mechanisms for participation and two-way dialogue to enhance the effectiveness of communication (Chess, 2000; Kasperson, 1986; National Research Council, 1989). In the context of bioterrorism, engaging the public as allies and partners is an essential ingredient for a successful response (Centers for Disease Control and Prevention, 2002; Covello, 2003; Covello et al., 2001; Sorenson, 2004; Working Group on Governance Dilemmas in Bioterrorism Response, 2004). Participation can be fostered through citizen advisory boards, formative research, community forums, and partnerships with community-based organizations.

Postal workers and public health professionals both described the desire for more participation in the response. One worker summed up this sentiment; "Listen to us. If we have an opinion, take our opinion. Listen to it. You never know what you can pick up from someone else."

Postal workers also had suggestions for increasing participation in the event of a future bioterrorism event. One worker offered this suggestion; "Through our unions, if they . . . can make an advisory board of postal workers and the Health Department working with the postal employees." Another suggested that postal workers could serve as advisors to increase the extent to which public health professionals understand the postal environment. Many participants discussed the need for postal workers to be involved in the distribution of information. Noting there was little trust in information from postal management, one worker asserted, "You tend to trust when you have someone from your group of people, be it race, be it from your job. You have a tendency to trust them more." They suggested that public health professionals conduct training and provide resource information to enable workers on the floor to effectively share accurate information with coworkers.

IMPLICATIONS FOR PRACTICE AND RESEARCH

You never know when something is going to come through the mail . . . you think about it everyday...what's going to happen next.

This quotation from a postal worker brings home the urgency associated with the development of crisis and emergency risk communication. We have listened to some of those who lived through the anthrax attack to help shape a proposed conceptual model as practical guidance for CERC.

Paramount to effective CERC is the understanding that the process is two-way and nonlinear. We propose that the Social Amplification of Risk model (Fig. 2.1) serves as an umbrella under which race and class, the social

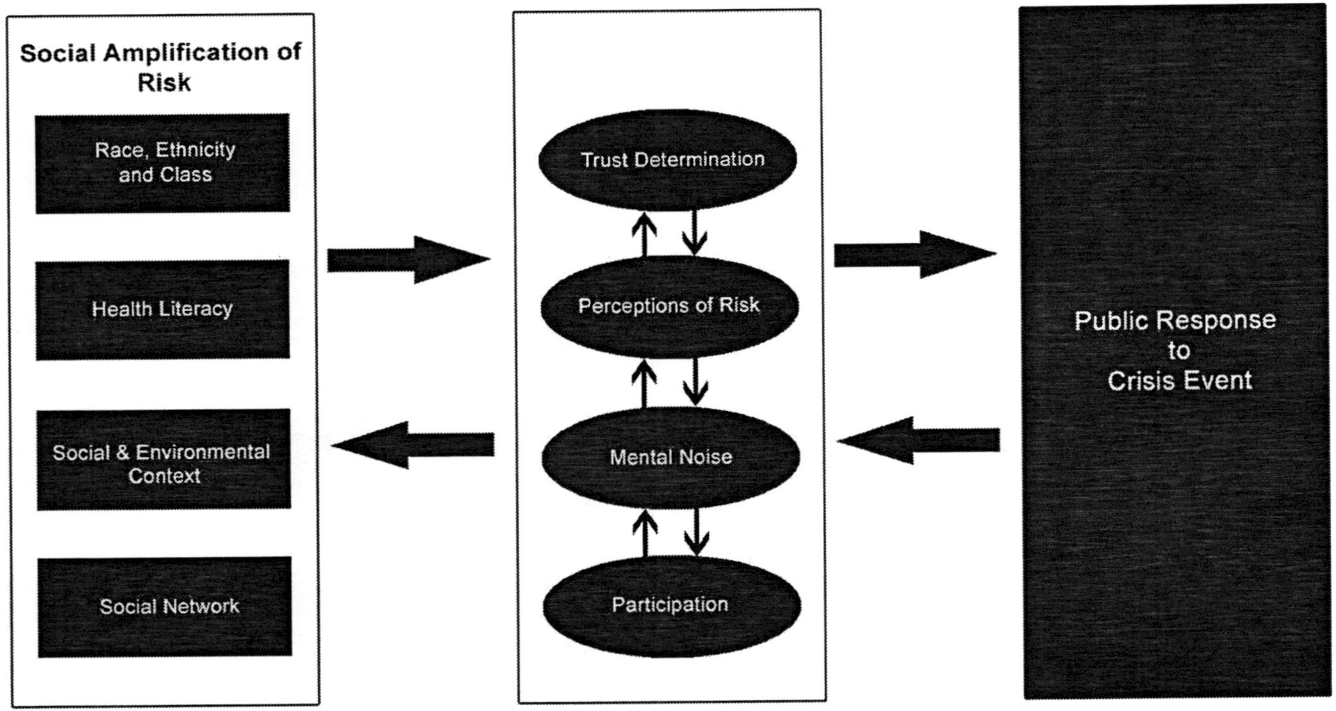

FIG. 2.1. A Proposed Conceptual Model for CERC.

network, health literacy, and the social and environmental contexts can be understood. Although we include them within the context of the social amplification of risk, we call attention to them because they were such meaningful components of the communication during the anthrax attack. We see these factors as both affecting and being affected by trust, participation, mental noise, and perceptions of risk. Finally, the response of the postal workers was ultimately shaped by these forces and, in turn, further altered the entire communication process.

For practitioners, the conceptual model provides a roadmap of factors to consider in the context of any CERC effort. However, determining the critical components of the conceptual model for any specific audience is a necessary part of any formative research phase. The primary challenge is that, in the context of CERC, there is often not adequate time to do extensive formative research. Therefore, the proposed model provides some guidance as to what may be considerations in any event. It must be supplemented with as much formative research with the affected audiences as is possible in the crisis situation. Formative research also increases opportunities for participation by the postal workers or any other pertinent audience.

The opportunities for research in the new field of CERC are significant, and certainly further work in the area of development and application of theories to CERC is critical. Although this conceptual model includes numerous factors that experience and previous research would suggest as valuable, evaluation of any future CERC effort would enable risk communicators to test the variables within the model to determine their utility with specific audiences and events and discover any new components that might be necessary for successful CERC.

The Working Group on Governance Dilemmas in Bioterrorism Response (2004) recognizes that "Each official action primes conditions for future public expectations and reactions" (p. 35). Furthermore, Sorenson (2004) questions a basic assumption of CERC; "The underlying premise is that if you issue the correct message through a credible spokesperson, people will respond in an appropriate manner to protect themselves" (p. 2). We advocate that the comprehensive and nuanced understanding of the dynamics of a crisis is essential to successful crisis and emergency risk communication. Consequently, it is critical that, as public health professionals, we examine the lessons from our first bioterrorist event and prepare ourselves to be more effective and trusted communicators in the future.

ACKNOWLEDGEMENTS

The authors wish to thank the postal workers and public health professionals who kindly shared their time and insights with us. This study was funded by Grant No. S2136-21/21 from the CDC/ASPH Cooperative Agreement. Dr. Quinn was also supported, in part, by the EXPORT Health Center at the Center for Minority Health, GSPH, University of Pittsburgh, NIH/NCMHD Grant No. 5P60MD000207-03.

REFERENCES

Blanchard, J., Haywood, Y., Stein, B., Tanielian, T., Stoto, M., & Lurie, N. (2004). In their own words: Lessons learned from those exposed to anthrax. *American Journal of Public Health, 95*(3), 489–495.

Centers for Disease Control and Prevention. (2001). Evaluation of bacillus anthrax contamination inside the Brentwood mail processing and distribution center—District of Columbia. *Morbidity and Mortality Weekly Report, 50*(50), 1129–1133.

Centers for Disease Control and Prevention. (2002a). *Crisis and emergency risk communication.*

Centers for Disease Control and Prevention. (2002b). Message mapping. In *Crisis and risk communication CDCynergy.*

Chess, C. (2000). Improving public participation in solving environmental health problems. *Journal of Environmental Health, 63*, 24–28.

Covello, V. (2003). Best practices in public health risk and crisis communication. *Journal of Health Communication, 8*, 5–8.

Covello, V., & Sandman, P. (2002). Risk communication: Evolution and revolution. In A. Wolbarst (Ed.), *Solutions to an environment in peril* (pp. 164–178). Baltimore, MD: Johns Hopkins University Press.

Covello, V., Peters, R., Wojtecki, J., & Hyde, R. (2001). Risk communication, the West Nile Virus epidemic, and bio-terrorism: Responding to the communication challenges posed by the intentional or unintentional release of a pathogen in an urban setting. *Journal of Urban Health, 78*(2), 382–391.

Earp, J., & Ennett, S. (1991). Conceptual models for health education research and practice. *Health Education Research, 6*(2), 163–171.

Fischoff, B. (2002). Assessing and communicating the risks of terrorism. In A. H. Teich, S. D. Nelson, & S. J. Lita (Eds.), *Science and technology in a vulnerable world* (pp. 51–64). Washington, DC: AAAS.

Fischhoff, B. (in press). Realistic risk disclosure in newly normal times. In P. Shane (Ed.), *Security, technology and privacy*. New York: The Century Foundation.

Fischoff, B., Bostrom, A., & Quaqdrel, M. (1993). Risk perception and communication. *Annual Review of Public Health, 14*, 183–203.

Frewer, L. (2004). The public and effective risk communication. *Toxicology Letter, 149*, 391-397.

Government Accountability Office. (2004). *U.S. Postal Service: Better guidance is needed to ensure an appropriate response to anthrax contamination.* Retrieved September 30, 2004, from www.Gao.gov/cgi-bin/getrpt?GAO-04-239.

Greene, C. M., Reefhuis, J., Tan, C., Fiore, A. E., Goldstein, S., Beach, M. J. et al. (2002). Epidemiologic investigation of bioterrorism-related anthrax, New Jersey, 2001. *Emerging Infectious Diseases*, *8*(10), 1048–1055.

Jernigan, D. B., Raghunathan, P. L., Bell, B. P., Brechner, R., Bresnitz, E. A., Butler, J. C. et al. (2002). Investigation of bioterrorism-related anthrax, United States, 2001: Epidemiological findings. *Emerging Infectious Diseases*, *8*(2), 1019–1028.

Kasperson, R. (1986). Six propositions on public participation and their relevance for risk communication. *Risk Analysis*, *6*(3), 275–281.

Merriam-Webster Online Dictionary. Retrieved January 21, 2005, from www.mn.com/.

National Research Council. (1989). *Chapter 1: Introduction. Improving risk communication.* Washington, DC: National Academy Press.

Pidgeon, N., Henwood, K., & Maguire, K. (2001). Public health communication and the social amplification of risks: Present knowledge and future prospects. In P. Bennett & K. Calman (Eds.), *Risk communication and public health*. New York: Oxford University Press.

Quinn, S., Thomas, T., & McAllister, C. (2005). Postal workers' perspectives on communication during the 2001 anthrax attack. *Biosecurity and Bioterrorism: Biodefense Strategy, Practice and Science, 3*(3), 207-215.

Reynolds, B., & Seeger, M. (2005). Crisis and emergency risk communication as an integrative model. *Journal of Health Communication*, *10*, 43–55.

Scherer, C., & Cho, H. (2002). A social network contagion theory of risk perception. *Risk Analysis, 23*(2), 261–267.

Shore, D. (2003). Communicating in times of uncertainty: The need for trust. *Journal of Health Communication*, *8*, 13–14.

Slovic, P. (2002). Terrorism as hazard: A new species of trouble. *Risk Analysis*, *22*(3), 435–436.

Slovic P. (1999). Trust, emotion, sex, politics and science: Surveying the risk-assessment battlefield. *Risk Analysis, 19*(4), 689–701.

Sorenson, J. (2004). Risk communication and terrorism. *Biosecurity and Bioterrorism: Biodefense Strategy, Practice and Science, 2*(3), 1–3.

Stein, B., Tanielian, T., Ryan, G., Rhodes, H., Young, S., & Blanchard, J. (2005). A bitter pill to swallow: Nonadherence with prophylactic antibiotics during the anthrax attacks and the role of private physicians. *Biosecurity and Bioterrorism: Biodefense Strategy, Practice and Science,* *2*(3), 1–11.

Tan, C. G., Hardeep, S. S., Crawford, D. C., Redd, S. C., Beach, M. J., Buehler, J. W. et al. (2002). Surveillance for anthrax cases associated with contaminated letters, New Jersey, Delaware, and Pennsylvania, 2001. *Emerging Infectious Diseases*, *8*(10), 1073–1077.

Working Group on "Governance Dilemmas" in Bioterrorism Response. (2004). Leading during bioattacks and epidemics with the public's trust and help. *Biosecurity and Bioterrorism: Biodefense Strategy, Practice and Science*, *2*(1).

Wray, R., Kreuter, M., Jacobsen, H., Clements, B., & Evans, R. (2004). Theoretical perspectives for public communication preparedness for terrorist attacks. *Family and Community Health*, *27*(3), 232–241.

3

Tracking the Anthrax Story

Spokespersons and Effective Risk/Crisis Communication

Julie M. Novak

M. Scott Barrett

Since 9/11, public health organizations increasingly confront risks and crises related to bioterrorism and emerging infectious diseases. Print media coverage provides the communication artifacts for an analysis of the Centers for Disease Control's (CDC's) response to the anthrax crisis in 2001. The CDC's response reveals the ramifications of failing to develop, in advance, a crisis plan, and also failing to designate a primary spokesperson or specified and previously trained spokespersons who can frame the crisis and concurrently coordinate messages that acknowledge the complexity, ambiguity, and uncertainty of the event. As a consequence, the CDC's response served to undermine the public health agency's authority and credibility and exacerbate rather than mitigate the crisis. Public health organizations can profoundly affect the trajectory and outcomes of risk and crisis situations through effective crisis plans and communication by their spokesperson/s.

Public health professionals regularly communicate with diverse audiences about health risk information. This includes risks and preventive measures for alcohol, tobacco, and drug use; chronic diseases; environmental toxins; and prenatal and well-child health care. Since 9/11, risks related to bioterrorism and emerging infectious diseases have joined the myriad of health issues that affect individuals and the general public (Covello, 2003). Communicating about such potentially life-and-death issues, when done well, fosters trust and confidence among professionals, the public, and other stakeholders and actually assists with overall risk assessment and ongoing decision making (U.S. Department of Health and Human Services, 2002). When done poorly, risk communication may actually increase the risk by creating confusion, inhibiting information flow and decision coordination, and inciting outrage and panic (Covello, 1992; Covello, Peters, Wojtecki, & Hyde, 2001). Therefore, public health organizations can profoundly affect the trajectory and outcomes of risk and crisis situations through effective communication by their spokespersons.

Public health organizations inevitably confront *crisis* situations at some point in time (Cohn, 2000). Seeger, Sellnow, and Ulmer (1998) define crisis as a "specific, unexpected, and nonroutine event or series of events that create high levels of uncertainty and threaten or are perceived to threaten an organization's high-priority goals" (p. 233). For public health organizations, crisis communication like risk communication involves "the exchange of information among interested parties about the nature, magnitude, significance, or control of a risk" (Covello, 1992, p. 359). The risk, however, has become an event. Unlike risk communication, crisis communication finds the experts as well as the public wrestling with ambiguity and uncertainty about the nature, magnitude, significance, or control of an unfolding event (Renn, 2003). A primary purpose of crisis communication is to reduce public uncertainty and anxiety through the dissemination of timely and accurate information based on what is known and not known so that an informed public can subsequently act (Sellnow, Seeger, & Ulmer, 2002). Crisis management is "a set of anticipatory measures that enables an organization to coordinate and control its response to an emergency" (Nudell & Antokol, 1988, p. 21). Ideally, public health organizations respond to crises, in part, by implementing plans developed in advance that facilitate effective communication with the public and other stakeholders (Olaniran & Williams, 2001).

Crisis communication plans prepare organizations to better meet their obligations to all stakeholders during a crisis, manage crises and facilitate coordinated decision making, and minimize the damage such events can do to organizational image, reputation, and credibility (Fearn-Banks, 2002; Seeger, Sellnow, & Ulmer, 2003). Crisis planning entails "establishing and maintaining functional communication networks and relationships with important constituencies, developing crisis teams, specifying spokespersons, and establishing preset responses" (Seeger et al., 2003, p. 169). Coombs (1999) and the Federal Emergency Management Agency (n.d.) outline necessary components in crisis plans. Most crisis plans, and notably those retrospectively considered effective, encourage the appointment of a primary spokesperson to ensure consistent messages (Benoit, 1997; Coombs, 1999; Kaufmann, Kesner, & Hazen, 1994; Rugo, 2001b; Turner, 1999).

During a crisis, the spokesperson serves as an organization's public face, acting as the conduit of information and explaining the crisis to stakeholders, especially the public. Although the spokesperson may rely on technical experts for specific information, the spokesperson retains overall responsibility for communicating (Seeger et al., 2003). The spokesperson frames the crisis and communicates the "tone and initial trajectory of the organizational response" as well as the "sensemaking process that ultimately leads to coordinated, harm-reducing actions" (Seeger et al. 1998, pp. 239–240). Therefore, a formal leader of the organization usually is the optimal and most effective primary spokesperson. In the absence of a primary spokesperson, stakeholders may perceive the messages that emerge from the myriad of spokespersons as inconsistent, unorganized, and untrustworthy, thereby escalating a loss of public trust or confidence in the organization. Ultimately, the crisis may be made worse.

The CDC's response to the 2001 anthrax crisis as a case study underscores the inadequacy of risk communication models and strategies for crisis events. Crises surprise, threaten, and demand immediate reaction, often with partial and incomplete information. Nonetheless, stakeholders and the media demand fairly incessantly information and instantaneous updates. The CDC's response reveals the ramifications of failing to designate a primary spokesperson or specified and previously trained spokespersons who can frame the crisis and concurrently coordinate messages that acknowledge the complexity, ambiguity, and uncertainty of the event and also include stakeholders in the ongoing efforts to mitigate the consequences. The myriad of spokespersons in the CDC's anthrax crisis response, as identified in the print media, served to undermine the public health agency's authority and credibility and exacerbate rather than mitigate the crisis.

CDC'S RESPONSE TO THE ANTHRAX SCARE

On or soon after the terrorist attack of 9/11, multiple anthrax-laced letters were mailed in the United States. One was mailed to *The Sun*, a Boca Raton newspaper, and received by photo editor Bob Stevens. Later, similar letters were received by *The New York Post*, NBC anchor Tom Brokaw, and Senate

Majority Leader Tom Daschle. The extensive infrastructure of the postal system meant that many individuals, not linked by a common work site or organizational affiliation, became exposed to inhalation anthrax, a previously rare hazard in the United States. The CDC, as the premier national public health agency, confronted a potential crisis of unknown dimensions.

On October 4, 2001, the CDC in Atlanta, Georgia, released a statement that Bob Stevens of Florida had died of inhalation anthrax. This initial response by the CDC portrayed the unfortunate case as an isolated incident. After all, the United States had experienced only 18 documented cases of inhalation anthrax throughout the previous century. Within days, however, as the anthrax attack spread to the Capitol, the CDC acknowledged a full-fledged crisis. Ultimately, 22 cases of inhalation anthrax resulting from tainted letters were diagnosed, and five people died. By the time the crisis ended, the public and other stakeholders viewed the CDC as a slow and rigid organization with ineffective communication and serious credibility problems; the CDC became an organizational casualty of the anthrax-related crisis.

Print media coverage provides the communication artifacts that reveal the CDC's ineffective communication interface of spokespersons with the public and other stakeholders. Three online databases (Lexus-Nexus, Infotrac, and the Electric Library) collectively identified 503 anthrax- and CDC-related news stories that appeared in major U.S. newspapers from September 1, 2001, through February 25, 2002 (Barrett et al., 2004). The three databases were selected based on their comprehensive indexing of a wide variety of major newspapers throughout the United States. The search focused on the words *CDC* and *anthrax*.

Each news article was read to identify the presence or absence of a cited spokesperson. In those articles with a cited spokesperson, the individual was considered a CDC spokesperson when the article noted some affiliation with the CDC. Otherwise, the individual was not considered a CDC spokesperson. Throughout the crisis, three general types of cited spokespersons emerged in the newsprint media: (a) official and formal CDC spokespersons, including administrators of the CDC and official spokespersons with authorization to engage in informed communication about the anthrax crisis and the investigation; (b) unofficial CDC and HHS spokespersons, including employees without formal authorization to act as spokespersons; and (c) unofficial and informal spokespersons, including anyone called on by the press to comment, such as former CDC staff and non-CDC bioterrorist and anthrax experts.

The 503 news stories cited a variety of individuals as spokespersons for the CDC. By examining newspaper accounts of the crisis, Barrett et al. (2004) identified by name and position title 81 distinct individuals purportedly speaking on CDC's behalf. CDC Director Koplan and Secretary of

Health and Human Services (HHS) Tommy Thompson[1] were quoted most frequently, whereas 79 other representatives were quoted at least once, some as many as 41 times; 62 of the 81 were CDC employees. The media clearly accessed a large number of CDC employees, regardless of whether they were official spokespersons, and the result was less than ideal. Because so many individuals spoke freely, and apparently in many cases without the assistance or guidance of trained public relations personnel, the CDC was unable to manage well its risk and crisis communication.

Concern over the number of official and unofficial CDC spokespersons received front-page attention in *The San Francisco Chronicle*, which quoted Carole Gorney, a Lehigh University crisis management specialist, as saying, "Part of the problem is that information is coming out from several different sources." The uncontrolled number of spokespersons, compounded by the varying degrees of their knowledge and informed status about the specific crisis, increased the public's uncertainty and anxiety. As aptly noted in the article, "beyond the medical unknowns, many experts said inconsistent public pronouncements have added to the confusion about the danger of disease spreading" (Hall & Stannard, 2001, p. A1).

RISK/CRISIS COMMUNICATION AND SPOKESPERSONS

Although a public health organization cannot predict a crisis, it can develop in advance a crisis plan and implement strategies that prepare the organization and a network of stakeholders for an effective response when such an event occurs. Retrospective reviews of plans that led to successful responses highlight some defining characteristics. First, a public health organization should be willing to share information, rather than conceal information under the guise that only public health experts can discern the importance, relevance, and severity of information especially when the potential impact is uncertain and frightening (Marra, 1998). Second, to maintain legitimacy, a public health organization should be willing to acknowledge and accept responsibility for any perceived harmful mistakes throughout the ongoing crisis (Hearit, 2001; Sellnow, Ulmer, & Snider, 1998). Third, a public health organization must be flexible enough to meet the diverse needs of its various stakeholders (Benoit, 1997; Covello et al., 2001; Sellnow & Ulmer, 1995). Finally, a public health organization should implement strategies developed in advance for informing the public about the risks and possible actions that may reduce their personal risk (Heath & Abel, 1996a).

[1]The CDC is under the supervision of the Secretary of HHS. Hence, Secretary Thompson was empowered to speak on behalf of the CDC during the anthrax crisis.

By preparing a crisis plan and a network of interacting stakeholders prior to crisis events, a public health organization can quickly initiate and coordinate responses that appropriately engage the public and other stakeholders, effectively attending to diverse communication needs (Marra, 1998). When a public health organization actively communicates and exchanges information, the public and other stakeholders can assist with ongoing risk management and decision making (Covello, 1992; Osterholm & Schwartz, 2000; Seeger et al., 2001). Agencies and stakeholders assist in crisis management by identifying, requesting, and disseminating useful information, an inherently ambiguous qualification that can only be defined by those affected by the crisis. The development of multi-agency and stakeholder networks as part of a crisis plan creates essential channels of communication and facilitates interaction that subsequently improves cooperation and coordination during crises (Osterholm & Schwartz, 2000). The coordinated efforts help to mitigate and resolve crises more quickly (Seeger et al., 2001).

Organizational procedures established to meet the needs of the media are also a key component in crisis plans. In the case of a crisis with widespread public interest and impact, the press promptly seeks information. Intense media scrutiny and competition increase the likelihood that journalists will attempt to access organizational members of varying affiliation, jobs, and levels of knowledge. As seen with 9/11 and other acts of terrorism, the media provide unparalleled coverage; broadcast media follow the unfolding crises around the clock for many days and weeks while newsprint media report on the crises day after day after day. Journalists are likely to seek out any and all members of an organization who are willing to provide information (Covello, 1992). The intense and prolonged coverage of crises differs drastically from the planned and limited coverage related to risk, the type of coverage and communication more typically encountered and experienced by public health professionals. Public health organizations must now be prepared to accommodate the media when crisis events occur and as they unfold (Balian, 1999; Heath & Abel, 1996b).

Effective crisis plans specify a media component, which includes the designation of spokesperson/s, the identification of communication channels, and the mechanisms for message development and dispersion. Because of this, all levels of employees should be informed on the organization's general crisis plan and provisions (Turner, 1999). If the organization has multiple people releasing information, the message/s and time of release should be coordinated (Balian, 1999). To avoid a lapse in the stream of information, an organization's communication with all involved parties—employees, investors and shareholders, and the media—should be continuous. This approach to information sharing reduces the risk of miscommunication or confusion regarding the organization's intended messages (Burton, 1989; Turner, 1999). Coordination with other organizations that share a common general goal is as important as coordination within an organization. In risk situations, apparent disagreements between agencies can lead to public mistrust. Because different organizations may have different, even incompatible, agendas, one organization should be in charge during a crisis (Osterholm & Schwartz, 2000). Coordinating organizations need to work together prior to and from the early stages of a crisis to make sure the public receives clear, consistent messages.

An effective way for a public health organization to manage the messages it provides to the public is to have a single, designated spokesperson. Generally, one designated spokesperson, rather than multiple spokespersons, performs more effective crisis communication (Benoit, 1997; Kaufmann et al., 1994; Rugo, 2001b; Turner, 1999). A trained spokesperson better understands and responds to crises than multiple untrained ones. When organizations designate one spokesperson in a crisis plan, the organization is more likely to ensure adequate training and preparation of the individual, who subsequently will knowledgeably assess and construct a system conducive to effective crisis communication. One spokesperson facilitates better consistency in messages, thereby avoiding the ramifications of mixed and confusing messages (Kaufmann et al., 1994). Although crises by nature are complex, ambiguous, and uncertain, one spokesperson can better maintain message consistency by coordinating the many messages emanating from within the organization and the network of coordinating agencies and not overstate claims when much is unknown. When organizations rely on one designated spokesperson, over time the spokesperson personifies the response. Therefore, many crisis managers believe the designated spokesperson should be a leader or top administrator in an organization because a leader represents the overall organization and wields the authority and legitimate power to make significant decisions (Turner, 1999). In some situations, the top administrator may not always be the best person for the spokesperson role (Kaufmann et al., 1994). For example, top administrators who lack public speaking skills or public savvy may cause more harm than good through unintentional missteps. Whether a top administrator should serve as spokesperson depends on the severity of the crisis and the administrator's willingness to risk public scrutiny (Rugo, 2001a). At the very least, the spokesperson should demonstrate an understanding of the roles of the public health organization and the press and the varied pressures they face in an unfolding crisis (Balian, 1999; Murphy, 1996). Finally, when organizations designate one spokesperson, the ability to calm and reassure stakeholders increases. One identifiable spokesperson builds confidence and trust among stakeholders due to a heightened perception of the organization's willingness to accept accountability.

Effective spokespersons require individual skills, position clarity in responsibilities and authority, and organizational support. To act quickly as befitting in crises, a spokesperson must be an autonomous, experienced, and media-trained person who is prepared, informed, and self-controlled (Marra, 1998; Nicolazzo & Nickson, 2001; Rugo, 2001a). The key to public receptiveness of a spokesperson is credibility. Covello et al. (2001) note that not only do spokespersons need to be trustworthy; they must also be the best persons to communicate messages of risk and crisis response. The spokesperson often becomes the "organization's public face, playing a critical role by providing information and explaining the crisis to stakeholders and the larger public" (Seeger et al., 2003, p. 239). The spokesperson serves to frame the crisis, explain ongoing activities to investigate and mitigate the crisis, keep stakeholders and the public informed, convey a sense of calm and control, and provide options for individual actions. If more challenging information must be communicated with a sense of credibility, a technical expert can be trained on message delivery and supported by an experienced spokesperson (Heath, 1995).

LESSONS LEARNED IN THE CDC AND ANTHRAX CASE

The intense demand for spokespersons by the news media during the anthrax crisis overwhelmed the CDC's crisis preparedness and ability to adequately respond. Stakeholders received confusing and contradictory messages from a variety of spokespersons, many only loosely affiliated with the CDC. As a result, the public health system—specifically the CDC—undermined its own credibility. The *San Francisco Chronicle* characterized the national public health system's overall response, stating, "There has been no single, competent authority on the bioterror attacks at home. The CDC has appeared reluctant to step into the breach, allowing mixed messages to go through state and local health agencies" (Hall & Stannard, 2001, p. A1).

Unfortunately, the CDC had not developed a crisis communication plan that sufficiently prepared the organization for a coordinated intra-organizational or interagency public health response. The CDC struggled with the designation and management of spokespersons throughout the crisis. There was no apparent coordination among communication experts, subject matter experts, and authorized agency decision makers. Within and outside the agency, the systems through which communication messages were developed, sent, or received appeared to be overwhelmed. It is likely that the CDC simply could not respond to the volume of media requests.

Despite 26 telebriefings and 24 formal press releases, the media still sought more statements and interviews from the CDC, sometimes to no avail. The Pittsburgh *Post-Gazette* implied as much when it commented, "CDC officials would not return numerous calls seeking comment" ("Labs told to Watch," 2001, n.p.). This suggested lack of responsiveness again reflected the inadequacies of the CDC's crisis communication systems, rather than any intentional disregard for information requests. Simply stated, the CDC had inadequate information capacity to fill an information vacuum created by the 9/11 attacks. In response to the intense and seemingly insatiable media demands for information and the CDC's perceived inability to meet such demands, independent spokespersons emerged to fill the informational void.

At times, external political authorities rather than public health experts or communication professionals appeared to arbitrarily decide who could respond to the press. The *Atlanta Journal and Constitution* reported on October 28, 2001, that "privately, CDC staff say they were kept from speaking out in the early days of the outbreak by orders from HHS" (McKenna, 2001, n.p.). Further, the *Boston Globe* reported, "In the first two weeks of the anthrax crisis, senior Bush administration officials told both Koplan [CDC's Director] and Surgeon General David Satcher to remain publicly silent" (Donnelly, 2001, n.p.). Although HHS Secretary Tommy Thompson had oversight responsibility for the CDC, he was neither a technical expert nor a communication expert. Therefore, in several cases, he appeared ill equipped, unprepared, and even confused in responding to questions as a principal CDC spokesperson. Some of his remarks were inaccurate or misleading, a fact quickly identified and chronicled by the media. For example, he commented that the nation should be on the lookout for "mysterious health symptoms," a comment roundly criticized by the media for inciting unnecessary anxiety and panic (Morse & Stoltz, 2001).

Moreover, the anthrax crisis brought unparalleled dimensions of ambiguity, uncertainty, and complexity that quickly revealed weaknesses of risk communication strategies traditionally utilized by public health organizations. The anthrax crisis was ambiguous and uncertain because little was known about inhalation anthrax and the consequences of potentially widespread exposure. The *Atlanta Journal and Constitution* reported, "When Bob Stevens . . . died Oct. 5 from inhaled anthrax, it was the first fatal U.S. case since 1976. Prior to that, only 18 cases of inhalational anthrax had been seen in this country during the past 100 years" (Seabrook, 2001, n.p.). The CDC, accustomed to providing conclusive expert knowledge, found themselves unable to provide such information to a demanding public. Unfortunately, the CDC did not initially discuss the uncertain and equivocal limits of their scientific knowledge. Rather, they asserted, "the inhaled form of anthrax could not be contracted through sealed letters" (Borenstein, Murphy, & Pugh, 2001, n.p.). "[CDC] Director Jeffrey Koplan said it was 'highly unlikely to virtually impossible' for someone to develop pulmonary anthrax from spores that floated from one piece of mail to another" (Connolly & Nakashima, 2001, n.p.).

By stating its position in such absolute and unequivocal terms, the CDC may have temporarily calmed the fears of millions of Americans, but at a high cost. When more people exposed to anthrax through the mail became sick, the CDC's reputation and credibility suffered greatly. Senator Tom Harkin was quoted as being "upset because he had thought the CDC 'was really on top of this' and it wasn't" (Borenstein et al., 2001, n.p.). It was widely reported that Harkin told CDC Director Koplan, "Maybe I'm wrong, but it just seems to me that something broke down here or is broken down. It's obvious people are getting sick, people are dying, and we can't afford to keep letting this happen. . . . I am very concerned about what CDC is doing and how they are operating" (McClam, 2001, n.p.; see also, McKenna, 2001). Similarly, health professionals questioned the credibility of CDC recommendations. The *Washington Post* reported that Health Commissioner George DiFerdinando acted against CDC advice when, on October 19, he "instructed all 1,000 postal workers at the Hamilton processing center to begin taking antibiotics as a precaution" (Russakoff, 2001, n.p.). Said DiFerdinando, "Epidemiologically, I had no data. This was a gut decision, and it's the decision I'm proudest of" (Russakoff, 2001, n.p.). Within three days, two postal workers "would die of pulmonary anthrax and [CDC] health officials would put thousands of District postal workers on antibiotics" (Russakoff, 2001, n.p.).

Simply stated, the anthrax crisis was new territory for public health professionals. Precise risk probabilities and predictions could not be made about consequences from weaponized anthrax deliveries through the mail system. Regardless of this uncertainty, CDC spokespersons proceeded to overreassure stakeholders and the general public. Moreover, when the CDC communication system was overwhelmed, the agency lost additional control over the message. The result was heightened confusion, increased public anxiety, and diminished credibility.

The crisis was also complex because of the context in which it occurred. The United States had only begun to emerge from the shock of 9/11: The nation was reeling in the aftermath and fearful of more terrorist attacks. After one supposedly isolated death from inhalation anthrax, the anthrax scare evolved to more cases and more deaths in Florida, New Jersey, New York, Washington, DC, and Connecticut. Cross-contaminated mail processed through the Brentwood, New Jersey, postal facility had traveled to dozens of locations throughout the country. The anthrax crisis was embedded within the 9/11 crisis, and this influenced the reception and the interpretation of every public health message.

In summary, postcrisis assessment of the CDC's response to the anthrax crisis reveals multiple areas for organizational learning and changes. First, public health organizations need to understand the distinct roles, styles, and content of risk communication and crisis communication. Strategies employed to inform stakeholders about risk and preventive measures differ from those effective during crisis events. Second, public health organizations need to develop crisis plans prior to crisis events so that coordinated responses can begin at the split-second onset of crisis. Although crises will occur, the exact timing or specifics of a crisis are by definition unknown in advance. Third, public health organizations need to designate the spokesperson/s before an event and ensure appropriate training. Although one spokesperson is often best, crises such as the anthrax crisis may require multiple spokespersons. In addition to training on crisis communication, training on roles and coordination are needed when there are multiple spokespersons. Fourth, as part of the crisis plan, public health organizations need to clarify the roles of cooperating agencies and interagency coordination. Part of the perceived confusion in the anthrax crisis stemmed from the lack of coordination between the CDC and HHS. Fifth, public health organizations need to develop a system for communication with surge capacity. Given that crises differ in magnitude, complexity, ambiguity, and uncertainty, the demands for information vary. In crises like the anthrax crisis, the CDC would have achieved better results had they developed capacity in advance to add additional spokespersons and channels of communication without losing coordination. Finally, public health organizations may better meet the demands for information by the media if the communication system engages in a sort of triage, routing requests by content and urgency to appropriate professionals.

CONCLUSIONS

The CDC exhibited inadequate crisis communication during the anthrax crisis. The intense media coverage revealed the CDC as ill prepared to coordinate a national public health response through crisis communication. In the absence of a well-developed and disseminated crisis plan, the public health agency relied on inappropriate and inadequate yet familiar strategies of risk communication. Notable was the unwieldy number of spokespersons. With no clearly designated spokesperson/s, large numbers of individuals attempted to respond to the public's clamor for information, although many had limited access to relevant knowledge. With so many different spokespersons speaking on behalf of the CDC, officially and unofficially, the CDC did not effectively convey its messages to the public and other stakeholders. The spokespersons initially sent messages with conclusive statements before such statements were prudent. Follow-up messages seemed to confuse and contradict rather than clarify.

Since 9/11, the CDC has engaged in postcrisis learning by identifying differences between risk and crisis communication and developing crisis communication systems to better prepare the organization and official spokespersons for future crisis situations. Previous research attests that proper crisis planning is essential and is characterized by the establishment of crisis communication systems before crises emerge. Such systems should be fashioned after historically successful plans and should include mechanisms allowing for continuous communication with stakeholders through the media. Although all public health personnel must be aware of the general provisions in a crisis plan, spokespersons must play the pivotal role in coordinating, sending, and receiving messages.

Effective crisis communication by designated and trained spokespersons helps public health organizations maintain trust and credibility. Throughout crises, public health organizations must exhibit willingness to promptly distribute information and avoid secrecy, readiness to admit fault as necessary and accept responsibility, and flexibility to meet the diverse needs of all stakeholders. These characteristics are especially vital in ambiguous, uncertain, and complex situations.

Moreover, the anthrax crisis and the CDC's response demonstrate the importance of establishing procedures to continuously meet media needs, training spokespersons to provide unified messages at appropriate times, training senior staff and all employees on the crisis plan, and presenting centralized and consistent messages. Although one spokesperson is ideal, public health organizations may encounter crises that require complex responses and multiple spokespersons. The horror of 9/11 and the anthrax crisis affirms the importance of preparedness. Crises have and will occur again; therefore, crisis plans need to guide public health organizations in appropriate preparation, including the development of communication systems and trained spokespersons.

REFERENCES

Balian, J. (1999). It's always an emergency with the media. *The Safety and Health Practitioner, 17,* 68–72.

Barrett, M. S., Berg, M. P., Grand, S., Hasbargen, K., Markey, V., Ocana, A. et al. (2004). The CDC and anthrax: The role of spokesperson in ambiguous and complex crises. *Speaker & Gavel,* 63–74.

Benoit, W. L. (1997). Image repair discourse and crisis communication. *Public Relations Review, 23,* 177–187.

Borenstein, S., Murphy, K., & Pugh, T. (2001, October 23). CDC criticized for response to anthrax cases. *Knight-Ridder/Tribune News Service.* Retrieved February 26, 2002, from http://www.elibrary.com.

Burton, R. M. (1989). Putting out forest fires. *Physician Executive, 15,* 22–23.

Connolly, C., & Nakashima, E. (2001, December 6). CDC sets "tips" on handling mail; precautions aimed at cutting risk from anthrax traces. *The Washington Post*. Retrieved February 26, 2002, from http://www.elibrary.com.

Cohn, R. (2000). *The PR crisis bible*. New York: St. Martins Press.

Coombs, W. T. (1999). *Ongoing crisis communication: Planning, managing, and responding*. Thousand Oaks, CA: Sage.

Covello, V. T. (1992). Risk communication: An emerging area of health communication and research. In S. A. Deetz (Ed.), *Communication Yearbook 15* (pp. 359–373). Newbury Park, CA: Sage.

Covello, V. T. (2003). Best practices in public health risk and crisis communication. *Journal of Health Communication, 8*, 5–8.

Covello, V. T., Peters, R. G., Wojtecki, J. G., & Hyde, R. C. (2001). Risk communication, the West Nile Virus epidemic, and bioterrorism: Responding to the communication challenges posed by the intentional or unintentional release of a pathogen in an urban setting. *Journal of Urban Health: Bulletin of the New York Academy of Medicine, 78*, 382–391.

Donnelly, J. (2001, November 11). *The Boston Globe*. Retrieved February 25, 2002, from http://web.lexis-nexis.com.

Fearn-Banks, K. (2002). *Crisis communications: A casebook approach* (2nd ed.). Mahwah, NJ: Erlbaum.

Federal Emergency Management Agency. (n.d.). *Emergency management guide for business and industry.* Retrieved April 13, 2005, from http://www.fema.gov/library/prepandprev.shtm#plan.

Hall, C., & Stannard, M. B. (2001, October 24). Anthrax scare strikes close to White House, lawmakers critical of CDC's reaction. *San Francisco Chronicle.* Retrieved December 9, 2001, from http://www.sfgate.com.

Hearit, K. M. (2001). When an organization speaks in defense of itself. In R. L. Heath (Ed.), *Handbook of public relations* (pp. 501–511). Thousand Oaks, CA: Sage.

Heath, R. L. (1995). Corporate environmental risk communication: Cases and practices along the Texas Gulf Coast. In B. R. Burleson (Ed.), *Communication Yearbook 18* (pp. 255–277). Thousand Oaks, CA: Sage.

Heath, R. L., & Abel, D. D. (1996a). Proactive response to citizen risk concerns: Increasing citizens' knowledge of emergency response practices. *Journal of Public Relations Research, 8*, 151–171.

Heath, R. L., & Abel, D. D. (1996b). Types of knowledge as predictors of company support: The role of information in risk communication. *Journal of Public Relations Research, 8*, 35–55.

Kaufmann, J., Kesner, I. F., & Hazen, T. L. (1994). The myth of full disclosure: A look at organizational communications during crises. *Business Horizons, 37*, 29–40.

Labs told to watch hazardous materials, CDC says verify inventory and safeguard shipping. (2001, October 17). *Pittsburgh Post-Gazette*. Retrieved February 26, 2002, from http://www.elibrary.com.

Marra, F. J. (1998). Crisis communication plans: Poor predictors of excellent crisis public relations. *Public Relations Review, 24*, 461–475.

McClam, E. (2001, November 29). Anthrax cases send dozens of CDC investigators across the country on secret missions. *AP Worldstream.* Retrieved December 7, 2002, from http://www.elibrary.com.

McKenna, M. A. J. (2001, October 28). America responds: At CDC, everything rearranged to put anthrax fight first. *The Atlanta Journal and Constitution*. Retrieved December 7, 2002, from http://www.elibrary.com.

Morse, S., & Stoltz, C. (2001, October 9). Symptoms include confusion. *The Washington Post*. Retrieved December 7, 2002, from http://www.elibrary.com.

Murphy, P. (1996). Chaos theory as a model for managing issues and crises. *Public Relations Review, 22*, 95–114.

Nicolazzo, R. E., & Nickson, S. (2001). The power of public opinion. *Risk Management, 48*, 41-47.

Olaniran, B. A., & Williams, D. E. (2001). Anticipatory model of crisis management: A vigilant response to technological crises. In R. L. Heath (Ed.), *Handbook of public relations* (pp. 487–500). Thousand Oaks, CA: Sage.

Osterholm, M. T., & Schwartz, J. (2000). *Living terrors.* New York: Dell.

Renn, O. (2003). Acrylamide: Lessons for risk management and communication. *Journal of Health Communication, 8*, 435–441.

Rugo, V. (2001a). Have your ducks in a row before a crisis. *Waste Age, 32*(6), 42–43.

Rugo, V. (2001b). How to communicate and prepare for a crisis. *Waste Age, 32*(3), 20.

Russakoff, D. (2001, November 11). Success by shoe leather, intuition; no new cases recorded since Oct. 31 signals apparent end to outbreak in state. *The Washington Post*. Retrieved February 26, 2002, from http://www.elibrary.com.

Seabrook, C. (2001, October 17). Flulike symptoms? It could be the flu; anthrax's rarity offers faint solace to fearful. *The Atlanta Journal and Constitution*. Retrieved December 7, 2002, from http://www.elibrary.com.

Seeger, M. W., Sellnow, T. L., & Ulmer, R. R. (1998). Communication, organization, and crisis. In M. E. Roloff (Ed.), *Communication Yearbook* (Vol. 21, pp. 231–275). Thousand Oaks, CA: Sage.

Seeger, M. W., Sellnow, T. L., & Ulmer, R. R. (2001). Public relations and crisis communication. In R. L. Heath (Ed.), *Handbook of public relations* (pp. 155–165). Thousand Oaks, CA: Sage.

Seeger, M. W., Sellnow, T. L., & Ulmer, R. R. (2003). *Communication and organizational crisis.* Westport, CT: Praeger.

Sellnow, T. L., Seeger, M. W., & Ulmer, R. R. (2002). Chaos theory, informational needs, and natural disasters. *Journal of Applied Communication Research, 30*, 269–292.

Sellnow, T. L., & Ulmer, R. R. (1995). Ambiguous argument as advocacy in organizational crisis communication. *Argumentation and Advocacy, 31*, 138–151.

Sellnow, T. L., Ulmer, R. R., & Snider, M. (1998). The compatibility of corrective action in organizational crisis communication. *Communication Quarterly, 46*, 60–75.

Turner, P. (1999). When the talking gets tough. *Journal of Property Management, 64*, 30–34.

U.S. Department of Health and Human Services. (2002). *Communicating in a crisis: Risk communication guidelines for public officials* (Reference Document No. SMA 02-3641). Washington, DC: Author.

PART III

Food-Borne Illness

4

Insider Agroterrorism Threat And High Reliability in a Turkey Processing Plant

Steven J. Venette

Agnes Ngale Lyonga

William E. Nganje

Timothy L. Sellnow

In 1984, a cult group poisoned salad bars with *Salmonella Typhimurium* bacteria at several Oregon restaurants causing an outbreak of salmonellosis. As a result, 751 people became ill. This act was described by federal agencies as the first recorded event of agroterrorism in the United States. The events of September 11, 2001, and the subsequent anthrax incidents increased concerns about unconventional terrorist attacks, including attacks on the U.S. food supply. This study uses survey data, cross-tabulations, and a discrete choice model to characterize risk within work groups and to determine the marginal effects of factors affecting the probability of insider threat in turkey processing. Processed turkey is a valuable food commodity, with a total U.S. production of $2.72 billion in 2003. The results indicate that being aware of rules preventing employees from carrying potentially dangerous materials onto the work floor, punishments against acts of intentional contamination, strictly following instructions to keep food safe and secure at work, and functioning as a High-Reliability Organization were significant determinants in minimizing insider threats. These results provide guidance to facilitate policies to sustain safer food.

The United States is the world's largest producer of food and agriculture products. The U.S. agriculture industry is a $200 billion business with over $55 billion in exports each year. Over 500,000 farms and 6,000 meat, poultry, and egg production establishments, including 57,000 food processors provide foods to local and export markets (U.S. Food and Drug Administration, 2004). Any deliberate act of sabotage or contamination in the U.S. food industry would be devastating to the economy and to public health (Pate & Cameron, 2001). Unfortunately, food-processing plants with widespread distribution networks, and employees from various countries who speak a variety of languages, remain highly vulnerable to deliberate disruption. According to the World Health Organization (2002), "Food is one of the most vulnerable sectors to intentional contamination by debilitating or lethal agents" (p. 6). Therefore, the contamination of food by terrorists poses a real and current threat.

The goal of the study presented in this chapter is not only to help increase awareness of food security, but also to identify or determine best practices for the industry by evaluating the applicability of high-reliability principles and risk communication strategies in turkey processing. To meet this goal, a case study of a turkey processing plant that is representative of a typical food-processing facility is examined. Turkey is an important food commodity, whose total value of U.S. production amounted to $2.72 billion in 2003 (Rawson, 2003). The plant examined in this case study has been operating since the 1930s, and its products are sold in every state in the United States and to foreign countries. The results of this study can serve as examples to other food-processing plants.

Specific objectives for this investigation included:

1. To characterize risk groups and evaluate the role of a high-reliability organization (HRO) in minimizing insider threat in turkey processing.
2. To determine the probability of insider threat in a turkey processing facility and the marginal impact of an HRO in minimizing insider threat.

This study also tests the hypothesis that food-processing facilities with employees who do not speak English well, if at all, pose a high insider threat that can increase the probability of intentional food contamination. Also, a discrete choice model is used in this chapter to characterize risk within work groups and determine the marginal effects of HRO in mitigating insider threats in turkey processing. However, this study suggests that an effective tool that can be used by food processors to mitigate processing risk is to follow high reliability principles that encourage effective risk communication between frontline employees, management, and policymakers.

REVIEW OF CONCEPTS

The study of HROs seeks to discover how people organize for high performance in settings where the potential for error and disaster are overwhelming (Rawson, 2003). An example would be air traffic control. Any mistake could cause a crash and loss of lives; however, relatively few accidents do occur. HROs focus on developing a culture of safety and encouraging mindful actions in organizations by minimizing errors and mitigating risks. This safety culture can reflect a highly reliable organization and may motivate organizational identification by employees.

Some organizations have learned that safety and risk mitigation are best developed in a culture that supports frontline workers and in which timely communication is encouraged to minimize human errors. Focus is placed on the right employees receiving the appropriate information at the correct time. HROs are able to use communication, training, and decision-making skill to minimize exposure to risk.

Risk has been defined by several authors as the probability of occurrence multiplied by the combination of hazard (in this case, the consequences of insider contamination) and outrage (how upset people would be if contamination were discovered; Covello, 1992; Sandman, 2000). A more complete definition explains that risk is a function of hazard, probability, control, and dread (Venette, 2003). The hazard is the potential negative event. Probability reflects the likelihood of the occurrence of that bad event. Control represents the ability to prevent or mitigate the harm. Dread is the perception of how awful the event would be if it did occur.

Risk communication focuses on the identification of threats, assessments and projections of possible harm, as well as management strategies to eliminate, reduce, and control the threats (Venette, 2003). Some risk scholars see risk communication as simply the process of informing the public of conclusions reached by risk analysts and managers. Others see risk communication as an integral part of both assessment and management processes, thus reflecting a complete risk analysis framework (Novak, 2004). Therefore, risk communication in this study is viewed as a context where all stakeholders—

those who are exposed to danger as much as those who generate it—should have some control over its management, as in an HRO culture.

For this chapter, the analyzed hazard comes from acts of food sabotage, such as the introduction of physical particles, microbial pathogens, and chemical or toxin agents at various points along the farm-to-table food continuum. Terrorists could attack livestock or crops to disrupt the food production system by deliberately contaminating food for human consumption with chemical, biological, or radionuclear agents for the purpose of causing injury or death to civilian populations or disrupting social, economic, or political stability (Rowan, 1991). Such an attack, according to the 2003 Government Accountability Office (GAO) report on bioterrorism, would cause severe disruption of the U.S. economy, because the agriculture sector accounts for 13% of the U.S. gross domestic product (GDP) and 18% of domestic employment. Terrorists could also contaminate processed food products if their primary motive is to cause bodily harm to humans.

The vulnerability of the food supply to potential acts of deliberate contamination is a national and international concern. Whether terrorists target food products or livestock and crops for deliberate contamination, serious public health and economic consequences are at stake. In today's global market, the contamination of food in one country can have a significant effect on public health in other parts of the world (Khan, Swerdlow, & Juranek, 2002). The mere threat of such an attack could seriously undermine consumer confidence in the safety of the food-supply and destabilize export markets. For example, when terrorists threatened to release foot and mouth disease in New Zealand, that nation's export markets and economy as a whole were placed in jeopardy. The discovery of even a single contaminated food product could result in nations closing their borders to trade of that product (Venette, 2003).

The potential impact of contaminated food on human health from deliberate acts of sabotage can be inferred from reports of unintended food-borne disease outbreaks (Sobel, Khan, & Swerdlow, 2002). A concerted, deliberate attack on food could be more devastating than an accident, especially if a powerful chemical, biological, or radionuclear agent is used. A terrorist using the food supply as a vehicle for attack would likely attempt to use an agent that would maximize the damage and deaths associated with the contamination.

Although a widespread terrorist attack on the American food industry has not occurred, examples of deliberate food contamination in the United States do exist. They include:

1. In September 1984, members of a religious cult contaminated salad bars in Dallas, Oregon, restaurants with *Salmonella typhimurium*, and 751 people became sick. This attack was reportedly a trial run

for a more extensive attack that was planned to disrupt local elections later that year (Torok, Tauxe, Wise, & Livengood, 1997).

2. In 1996, a reference strain of *Shigella dysenteriae* type 2 was used by a disgruntled laboratory worker to deliberately contaminate food to be consumed by colleagues. This contamination caused 12 people to become ill (Kolavic, Kimura, Simons, & Slutsker, 1997).
3. In January 2003, the U.S. Centers for Disease Control and Prevention reported that 92 people became ill after purchasing ground beef from a Michigan supermarket. The beef was intentionally contaminated with nicotine. An employee was indicted for intentionally poisoning 200 pounds of meat sold in the supermarket (Prichard, 2003).
4. In Irvine, California, on July 28, 2004, ground-up castor beans with trace amounts of ricin poison were found in two jars of baby food. Notes were also included that the food had been contaminated (Cable News Network, 2004).

RESEARCH QUESTIONS

The overarching research question in this study is: What role does risk communication in an HRO culture play in minimizing insider threats or deliberate acts of food contamination? Based on the review of relevant literature, three specific research questions were established for this study:

RQ_1 = What are the high-risk groups of employees in the case study plant based on the job title?
RQ_2 = Do employees at the case study plant perceive the plant to be an HRO?
RQ_3 = Is there low or high probability of intentional contamination in the processing plant, and what factors contribute to these risk levels?

METHODS

This study uses survey data, cross-tabulations, and a discrete choice model (binary logit) to characterize risk groups and determine the probability of low- or high-intentional contamination in turkey processing. The marginal effects of having characteristics of an HRO and other factors that mitigate the risks of insider agroterrorism threats are also determined. A survey questionnaire, available in English and Spanish, was distributed to 271 employees of a Midwest turkey plant. Of the 271 surveys sent to employees, 190 were completed and returned, yielding a response rate of 70%. Of the 190 surveys completed and returned, 167 were filled out in English, and 23 were filled out in Spanish. The survey consisted of 32 questions: 10 background and demographic questions, 12 questions about intentional contamination, and 10 HRO perception index questions. The demographic questions asked respondents to indicate job categories, primary language, nation of origin, and length of time they have worked in this and other processing plants. Chronbach's alpha for the survey instrument was .87.

The intentional contamination questions dealt with the possible ways that employees can identify areas and pathways of an insider threat. The HRO perception index and communication questions dealt with information sharing, the plant's management, concerns about the possibility of making a food safety error, the plant's commitment in correcting any shortcomings in the food safety inspection process, and whether employees' actions directly contribute to the prevention of food safety problems.

Cross-tabulations were used to characterize employees' risk groups based on job title. A discrete choice model (binary logit) was used to measure the probability of insider threat of intentional contamination in the case study processing plant. This method has been used successfully by several authors (Govindasamy, Hossain, & Adelaja, 2005; Nganje, Schuck, & Okigbo, 2005).

The goal of any processing firm is to minimize the probability of an outbreak from either unintentional or intentional contamination. The ability to minimize risk can be characterized as a discrete choice model (Nganje, Kaitibie, & Taban, 2005). In this case, the outcome is reduced to a logit model because of the two choices for the dependent variable (contamination $\gamma_i = 1$, or no contamination $\gamma_i = 0$). The dependent variable was proxied with questions that asked whether employees had experienced any incidence or act of deliberate contamination, had ever brought unwanted or restricted materials on the work floor, or had not followed work instructions. Table 4.1 summarizes the variables used in the logit model and their descriptive statistics.

The probability of deliberate contamination was measured by: $F(\chi\beta) = 1 / 1+e^{-\chi\beta}$ = (estimates) and $1 - F(\chi\beta)$ otherwise. A utility index function for contamination γ_i^* is calculated as $\gamma_i^* = \beta X + \varepsilon$, where X is a vector of independent or explanatory variables, i is the observation, and ε is the random error term.

Thus, the probability of contamination ($\gamma = 1$) is observed as a function of γ_i^* if $\gamma_i^* > 0$ or $X\beta + \varepsilon > 0$, and 0 otherwise. Marginal effects are calculated as $\partial F(\chi\beta) / \partial\chi\beta * \partial\chi\beta = f(\chi\beta) / \partial\chi * \beta j$ where $f(\chi\beta)$ is a probability density function equal to $e^{\chi\beta} / (1+e^{\chi\beta})^2 * \beta j$, and e are the natural log notation.

TABLE 4.1.
Variables and Their Descriptive Statistics Used in the Logit Model

VARIABLE DESCRIPTION	RESPONSES	DESCRIPTIVE STATISTICS		
		MEAN	MODE	SD
Do employees strictly follow instructions to keep food safe at work?	Always = 1 Many times = 2 Sometimes = 3 Not always = 4	2.02	1	2.34
Are employees aware of rules preventing them from carrying unwanted materials onto the work floor?	Yes = 1 No = 0	0.91	1	0.92
Are employees screened for unwanted materials before they begin work?	Yes = 1 No = 0	0.45	0	1.35
Do employees receive enough training to identify and tell acts of (intentional) contamination?	Strongly agree = 1 Agree = 2 Neither = 3 Disagree = 4 Strongly disagree = 5	2.91	2	1.35
Are employees aware of punishments against acts of deliberate food contamination?	Yes = 1 No = 0	0.87	1	1.42

TABLE 4.1.
Variables and Their Descriptive Statistics Used in the Logit Model *(continued)*

VARIABLE DESCRIPTION	RESPONSES	DESCRIPTIVE STATISTICS		
		MEAN	MODE	SD
Employees' nationality or state of origin?	SD = 1; ND = 2 MN = 3; TX=4 Mexico = 5 Vietnam = 6 Others = 7			
The plant is concerned about the possibility of making a food safety error (HRO Perception Index 1)	Strongly agree = 1 Agree = 2 Neither = 3 Disagree = 4 Strongly disagree = 5	2.22	2	1.06
The plant is committed to correcting any shortcomings in food safety inspection process (HRO Perception Index 2)	Strongly agree = 1 Agree = 2 Neither = 3 Disagree = 4 Strongly disagree = 5	2.11	2	0.94
Employees' actions directly contribute to the prevention of food safety problems (HRO Perception Index 3)	Strongly agree = 1 Agree = 2 Neither = 3 Disagree = 4 Strongly disagree = 5	2.26	2	1.03

RESULTS

> *RQ_1 = What are the high risk groups of employees in the case study plant based on the job title?*

The results of the cross-tabulations (Tables 4.2, 4.3, and 4.4) indicate that the categories of employees who are most likely to be considered higher risk groups are employees who work on the line, workers whose primary language is not English, and people who have emigrated to America. These categories represent employees who revealed that they generally are unaware of rules preventing them from carrying unwanted materials onto the work floor, are not aware of punishments against acts of deliberate food contamination, and sometimes or not always strictly follow instructions to keep food safe at work. Thus, they are characterized as a higher risk group.

Table 4.2 shows the categories of employees who strictly follow work instructions to keep food safe and those who do not. All of the managers strictly follow instructions to keep food safe at work, but only 17 of 22 supervisors, 99 of the 141 line workers, and 13 of the 18 janitorial workers strictly follow orders to keep food safe at work. Although six line workers sometimes strictly follow orders to keep food safe at work, five line workers do not always strictly follow orders to keep food safe, thereby also characterizing them as a higher risk group.

Table 4.3 presents a summary of employees who are aware of rules that prevent them from carrying unwanted materials onto the work floor. Eighty-two percent of employees in the case study plant are aware of rules preventing them from carrying unwanted materials onto the work floor, following the standard sanitation operating procedures, and handling of food safely. Again, the line workers are shown to be a high- risk group because 27 employees who directly handled and processed the product in various work areas (processing points) responded that they were not aware of rules banning unwanted materials from the work floor.

Table 4.4 summarizes the categories and numbers of employees who are aware of punishments against acts of intentional food contamination in the plant. Overall, only 63% of the employees are aware of punishments for acts of intentional contamination. Most important, 2 of the 3 managers are unaware of punishments for acts of deliberate food contamination, and 7 out of 22 supervisors are also not aware of these punishments.

> *RQ_2 = Do employees at the case study plant perceive the plant to be an HRO?*

Employees generally perceive their plant to be functioning as an HRO and this HRO structure of the plant has a high marginal impact on the low

TABLE 4.2.
Categories of Employees Who Strictly Follow Instructions to Keep Food Safe/Secure at Work

JOB TITLE	ALWAYS	MANY TIMES	SOMETIMES	FEW TIMES	NEVER	OMITTED	TOTAL
Management	3	0	0	0	0	0	3
Supervisor	17	3	1	0	0	1	22
Line worker	99	19	6	0	5	12	141
Office staff	4	1	0	0	0	0	5
Security	1	0	0	0	0	0	1
Cleaning	13	0	1	0	0	4	18
Total	137	23	8	0	5	17	190

TABLE 4.3.
Categories of Employees Who Are Aware of Rules Preventing Unwanted Materials at Work

JOB TITLE	UNWMA = 0*	UNWMA = 1	UNWMA = 9	TOTAL
JT = 1**	1	2	0	3
JT = 2	2	20	0	22
JT = 3	27	112	2	141
JT = 4	1	4	0	5
JT = 5	0	1	0	1
JT = 6	2	16	0	18
Total	33	155	2	190

*UNWMA = Are employees aware of rules preventing unwanted materials at work? UNWMA = 0 is no; UNWMA = 1 is yes; UNWMA = 9 is blank.
**JT (job title) = 1 is management; JT = 2 is supervisor; JT = 3 is line worker; JT = 4 is office staff; JT = 5 is security; and JT= 6 is cleaning and sanitization.

TABLE 4.4.
Categories of Employees Who Are Aware of Punishments Against Acts of Intentional Contamination

JOB TITLE	POIC = 0*	POIC = 1	POIC = 9	TOTAL
JT = 1**	2	1	0	3
JT = 2	7	15	0	22
JT = 3	53	84	4	141
JT = 4	0	4	1	5
JT = 5	0	1	0	1
JT = 6	3	15	0	18
Total	65	120	5	190

*POIC = Are employees aware of punishments against acts of intentional contamination? POIC = 0 is no; POIC = 1 yes; and POIC = 9 is blank.
**JT (job title) = 1 is management; JT = 2 is supervisor; JT = 3 is line worker; JT = 4 is office staff; JT = 5 is security; and JT = 6 is cleaning and sanitization.

probability of deliberate food contamination. This study indicates that some characteristics of HROs function in the turkey processing plant, which contributes significantly to the low probability of food sabotage in the workplace. The plant's concern about the possibility of making a food safety error and its commitment in correcting any shortcomings in the food safety inspection process are important HRO characteristics. Moreover, the data are consistent with a previous study that examined communication perceptions related to food safety risk at the case study plant, which suggested that employees generally perceived the plant to be an HRO (Novak, Sellnow, Venette, & Nganje, 2005).

Functioning as an HRO is a significant determinant that might lower the probability of intentional food contamination. All organizations face risk; however, high-reliability characteristics work to instill a culture of safety and mitigate the risk of failures in organizations. Therefore, risk-adverse organizations work to develop highly reliable practices by constantly surveying their surroundings for early signs of impending hazards. Consequently, one implication of functioning as an HRO is that employees detect and report any deviations from the normal processing procedures. The result of RQ_2 is validated with the marginal impacts from the following logit model.

RQ_3 = *Is there low or high probability of intentional contamination in the processing plant, and what factors contribute to these risk levels?*

The results predict that there is a low probability of an insider threat in this processing plant. The outcome of the discrete choice model revealed that the occurrence of intentional contamination measured by the dependent variable is low ($p = .01$); conversely, the probability of intentional contamination not occurring ($p = .85$) in the plant is high.

Table 4.5 presents the results of the binary logit model and the marginal effects of determinants of intentional contamination. Table 4.5 also summarizes the important determining factors affecting the probability of inten-

TABLE 4.5.
Parameter Estimates and Coefficients of Binary Logit Model of Employees' Characteristics on Intentional Contamination

VARIABLE NAMES	VARIABLE DESCRIPTION	COEFFICIENT	P	ELASTICITY
OVCPFS	HRO index 1	-0.06	.05	-1.16
s.e.		(0.03)		
OVCSFSI	HRO index 2	0.08	.01	1.44
s.e		(0.03)		
EACPFS	HRO index 3	-0.01	.67	-0.20
s.e		(0.02)		
FSINSTR	Work instructions	0.03	.00	0.51
s.e.		(0.01)		
UNWMAT	Unwanted materials	-0.08	.08	-0.61
s.e.		(0.04)		
ETIRAIC	Received training	-0.47	.01	-1.21
s.e.		(0.02)		
ORIGIN	Origin	-0.02	.02	-0.58
s.e.		(0.01)		
POIC	Punishments for IC	-0.00	.96	-0.01
s.e.		(0.02)		
McFadden R^2	.09			
Percentage Correct Predictions	86.32%			

Note. The values in parentheses are standard errors. Variables are defined in Table 4.1.

tional contamination. The coefficient, *p* value, elasticities, and standard errors for the binary logit model are included. The goodness-of-fit test shows a high predictive value of 86.32% for the model and a McFadden R^2 of .09. Several models were run, and the model was selected based on the goodness of fit, the number of significant variables, and the absence of multicollinearity among variables.

The majority of coefficients are significant. The first two HRO perception indexes (*the concern about the possibility of making a food safety error* and *the commitment to correct any shortcomings in food safety* are significant (p = .05 and .01, respectively), but the third HRO variable (*employees actions directly contribute to the prevention of food safety problems)* is not significant (p = .67). *Being aware of rules preventing employees from carrying unwanted materials onto the work floor* is also significant (p = .08). *Strictly following instructions at work to keep food safe and secure* is significant (p = .00), whereas *receiving enough training to keep food safe and secure by identifying and reporting any deviations from the normal processing practice* and *employees' nation of origin* are also both significant (p = .01 and .02, respectively). The results in Table 4.5 indicate that the HRO perception indexes (*strictly following work instructions, being aware of unwanted materials on the work floor, receiving enough training on food protection*, and *employees' nation of origin*) are significant determinants in minimizing insider food threats in the case study processing plant.

The marginal values (Table 4.6) show that the variables with least impact on the probability of no contamination (Y = 0) are *being aware of punish-*

TABLE 4.6.
Marginal Effects for Binary Logit Model

PARAMETERS	PROBABILITY OF Y = 0	PROBABILITY OF Y = 1	MARGINAL EFFECTS
1OVCPFS	-0.05	-0.10	-0.06
OVCSFSI	0.07	0.13	0.08
EACPFS	-0.01	-0.02	-0.01
FSINSTR	0.03	0.05	0.03
UNWMAT	-0.07	-0.13	-0.08
ETIRAIC	-0.04	-0.08	-0.05
ORIGIN	-0.02	-0.04	-0.02
POIC	-0.00	-0.00	-0.00

Note: Variables are defined in Table 4.1.

ments against acts of deliberate contamination and *employees' actions contributing to the prevention of food safety problems.* The variables with the greatest impact on the probability of no contamination (Y = 0) are *strictly following orders to keep food safe, commitment to correcting any shortcomings in food safety inspection, being aware of rules that prevent unwanted materials onto the work floor,* and *the plant's concern about the possibility of making a food safety error.* The data indicate that the HRO perception index is a significant determinant in minimizing the risk of insider threat or intentional contamination at the turkey processing facility.

CONCLUSIONS AND DISCUSSIONS

The binary choice model reveals a low probability of intentional food contamination in the case study processing plant. The data reveal that most employees not only perceive their organization as functioning as an HRO, but they also strictly followed orders to keep food safe at work. Employees' strict adherence to instructions suggests that the plant's management has been effective in communicating policy. Similarly, the data show that employees have developed a reliance on food safety information from top management officials who are mindful of delivering safe products to their customers.

Another indicator of high reliability is the finding that, although some categories of workers differ, overall most employees (81.6%) are aware of rules preventing them from carrying unwanted materials onto the work floor. Moreover, restricting access to unknown or unauthorized persons is already standard practice in the processing plant. Thus, the introduction of dangerous material is unlikely.

Unfortunately, not all of the data reflect effective policies or procedures. For example, employees did not perceive that they receive enough training on food protection and security. The marginal impact of this perception showed that if management increased training, the probability of contamination would decrease. Thus, increasing required training related to food protection and food safety risks, especially for newly hired employees, should be beneficial.

Overall, this research used a discrete choice model (binary logit) to reveal that the probability of intentional food contamination is low at the case study processing facility. But this finding should not be viewed as a reason to relax efforts to protect this sector of agriculture. Food still remains one of the most vulnerable sectors to intentional contamination by debilitating or lethal agents. Therefore, the contamination of food by terrorists poses a real threat. One lesson that the tragic events of September 11, 2001, have indelibly etched on the minds of most people is that terrorists will seek new and unexpected methods in their attempts to harm the United States (Casagrande, 2000). This nation's food supply and agricultural industries could be targets for these new methods. Although vulnerability cannot be completely eliminated, this study suggests the food-processing industry can minimize insider agroterrorism threats by functioning as an HRO, communicating food safety rules to all groups of employees, and targeting training needs to frontline and newly hired employees.

REFERENCES

Cable News Network. (2004). *FBI: Trace amounts of ricin found in baby food.* http://www.cnn.com/2004/US/West/07/28/poison.babyfood.ap/index.html.

Casagrande, R. (2000) Biological terrorism targeted at agriculture: The threat to US national security. *Nonproliferation Review, 7*, 98–99.

Chalk, P. (2001). Terrorism, infrastructure protection, and the US food and agricultural sector. *RAND, 184*(2), 1–11.

Covello, V. T. (1992). Risk communication: An emerging area of health communication research. In S. A. Deetz (Ed.), *Communication Yearbook 15* (pp. 359–373). Newbury Park, CA: Sage.

Govindasamy, R., Hossain, F., & Adelaja, A. (2005). *Income of farmers who use direct marketing.* Accessed March 14, 2005, from http://www.agecon.lib.umn. edu.

Joppen, L. (2003). Industry not prepared for 9/11 in food: WHO: "Food terror real and current threat." *Food Engineering & Ingredients, 28*, 30–35.

Kaneko, N., & Chern, W. S. (2005). *Consumer acceptance of genetically modified foods: A telephone survey.* Paper presented at the American Agricultural Economics Association meeting. Accessed March 14, 2005, from http://www.agecon.lib.umn.edu.

Khan, A. S., Swerdlow, D. L., & Juranek, D. D. (2002). Precautions against biological and chemical terrorism directed at food and water supplies. *Public Health Reports, 116*, 3–14.

Kolavic, S. A., Kimura, A., Simons, S. L., & Slutsker, L. (1997). An outbreak of *Shigella dysenteriae* type 2 among laboratory workers due to intentional food contamination. *The Journal of the American Medical Association, 278*(5), 396–399.

Nganje, W. E., Kaitibie, S., & Taban, T. (2005). Multinomial logit models comparing consumers and producers' risk perception of specialty meat. *International Journal of Agribusiness, 21*(3), 1–25.

Nganje, W. E., Schuck, E., & Okigbo, C. (2002). An economic analysis of information, education, and communication (IEC) in adoption of slash and burn agriculture. *Journal of African Communication, 4*, 136–158.

Nganje, W. E., Wilson, W. W., & Nolan, J. (2003). Agro-terrorism and the grain handling systems in Canada and the US. *Journal of Canada Agricultural Economics Society, 2*(200), 148–159.

Novak, J. M. (2004). *Using informant interviews to explore organizational culture in a turkey processing plant: How foregrounded communication impacts risk management.* Paper presented at the Red River Student Communication Conference, Fargo, ND.

Novak, J. M., Sellnow, T. L., Venette, S. J., & Nganje, W. E. (2005). Perceptions of food safety, risk, and high reliability: A case in turkey processing.

Pate, J., & Cameron, G. (2001). Covert biological weapons attacks against agricultural targets: Assessing the impact on US agriculture. *BCSIA Discussion Paper* (No. 2001-9).

Prichard, J. (2003). Ex-grocery worker sentenced to nine years for poisoning supermarket beef. *South Bend Tribune*. Accessed November 20, 2004, from http://southbendtribune.com/news/2003.html.

Rawson, J. M. (2003). Meat and poultry inspection issues. *Congressional Research Service Issue Brief for Congress* (Order Code IB10082), pp. 1–6.

Rowan, K. E. (1991). Goals, obstacles, and strategies in risk communication: A problem-solving approach to improving communication about risks. *Journal of Applied Communication Research, 19*, 300–329.

Sandman, P. M. (2000). *Open communication: Risk communication in food safety; Motivating and building trust.* Paper presented at the National Food Safety and Toxicology Center conference, East Lansing, MI.

Sobel, J., Khan, A. S., & Swerdlow, D. L. (2002). Threats of a biological terrorist attack on the US food supply: The CDC perspective. *The Lancet, 359*, 874–880.

Torok, T. J., Tauxe, R. V., Wise, R. P., & Livengood, J. R. (1997). A large community outbreak of Salmonellosis caused by intentional contamination of restaurant salad bars. *The Journal of the American Medical Association, 278*(5), 389–396.

U.S. Food and Drug Administration, Center for Food Safety & Applied Nutrition. (2004). *Risk assessment for food terrorism and other food safety concerns.* Accessed November 20, 2004, from http://www.cfsan.fda.gov/~dms/rabtact.html.

Venette, S. J. (2003). *Risk communication in a high reliability organization: APHIS PPQ's inclusion of risk in decision making.* Ann Arbor, MI: UMI Proquest Information and Learning.

Weick, K. E., & Sutcliffe K. M. (2001). *Managing the unexpected: Assuring high performance in an age of complexity.* San Francisco: Jossey-Bass.

World Health Organization, Food Safety Department. (2002). Terrorist threats to food, guidance for establishing and strengthening prevention and response systems. *Food Safety Issues, 28*(2), 5–16.

5

When Organizational Responses to a Health Crisis are Limited

Chi-Chi's and Hepatitis A Outbreak

Denise Gorsline

A November 2003 hepatitis A breakout at a Chi-Chi's restaurant in Monaco, Pennsylvania sickened over 600 people and killed 4. Precrisis organizational variables of Chi-Chi's poor financial performance and underdeveloped stakeholder relationships with the restaurant's suppliers presented insurmountable challenges. These conditions negatively affected the ability for Chi-Chi's to respond appropriately to the crisis and, ultimately, contributed to the company's dissolution.

Organizational crises are increasingly frequent occurrences, They are widely covered by the media, and they are closely examined by scholars from a variety of fields. The Institute for Crisis Management (2003) reports that there are 90,000 records of business crisis news stories since 1990. The instituted reported that 2001 saw the most crises since 1990, and the September 11 terrorist attacks were not the sole cause of the increase. The failing economy, the collapse of the dot.com industry, contaminated foods, product defects, strikes, and banking and security issues all represented significant crises in 2001 (Millar, 2003).

For at least the last 20 years, communication scholars have developed an increasing interest in organizational crises. Researchers have completed case studies on a wide variety of organizations and types of crises, including from retailers (Benoit, 1995), airlines (Ray, 1999), oil and investment conglomerates (Seeger & Ulmer, 2003), floods (Sellnow, Seeger, & Ulmer, 2002), medical crises (Johnson et al., 2004), to issues of food-borne illness (Ulmer & Sellnow, 2000). The purpose of these case studies has been to identify patterns of response to organizational crises and suggest practices for future events. This chapter examines what happens when these typical responses are not available to an organization facing a crisis.

CRISIS MANAGEMENT

With the growing scholarly and media attention, various patterns of crisis management have emerged. Researchers (Coombs, 1995; Pauchant & Mitroff, 1992; Ray, 1999; Turner, 1976) have examined the development of crisis in stages using a verity of different models (see Seeger, Sellnow, & Ulmer, 2003). Although the number of stages varies, these developmental approaches all identify a progression of crisis-related events and conditions. These approaches all describe a systematic movement from precrisis, or incubation period, to crisis, when harm accumulates, followed by post-crisis which included explanations and interpretations. These developmental stages have also been associated with specific actions or strategies that may be useful under those particular conditions.

The developmental approach to analyzing the effectiveness of communication during a crisis starts with the precrisis stage. In general, this includes an examination of the business processes, stakeholder relationships, preparation, and understanding of risk, which existed prior to the emergence of the crisis. The quality or style of leadership is one of the most important variables, including the ability to build positive stakeholder relationships.

Stakeholder relationships are critical to the success of an organization facing a crisis (Ulmer, 2001). These stakeholders include a wide variety of groups, all connected in some way to the organization, including, but not limited to, suppliers, stockholders, customers, and employees (Ulmer, 2001).

Although precrisis planning is advocated by almost all observers of crisis management, the exact form of the plan varies from organization to organization (Seeger, Sellnow, & Ulmer, 2003). Variations in form would not be a significant problem if the plans all existed. Regardless of the consistent endorsement of crisis planning as an important business practice, and although organization leaders believe crises are inevitable, many organizations do not have crisis plans that are robust and up to date (Seeger et al., 2003). The crisis stage is the shortest of the developmental stages, but is often characterized by high levels of uncertainty and confusion. The effects of the planning and business practices identified in the precrisis stage come to fruition when the organization is confronted with a crisis.

The postcrisis stage most frequently includes salvaging organizational legitimacy, organization learning, and healing. Salvaging organizational legitimacy may be the most significant goal, because the continued existence of the organization may depend on it (Seeger et al., 2003). The opportunity for organizational learning presented by a crisis goes beyond the daily monitoring and performance review that organizations routinely conduct. Healing must also take place if the organization and its stakeholders are to move beyond the disruption and harm of the crisis. Healing may mean renewal for the organization, or it may be merely relief that they survived a challenging time. In other cases, as with the case presented here, healing may be an unobtainable goal.

CHI-CHI'S

At the time of the hepatitis A breakout, Chi-Chi's was a chain of Mexican casual dining restaurants owned by Prandium, Inc. The chain consisted of 99 restaurants located in 17 states in the Midwest and East Coast regions of the United States. In November 2003, a Chi-Chi-s restaurant in Monaco, Pennsylvania, 25 miles from Pittsburgh, was involved in the nation's biggest outbreak of hepatitis A from a single source. This outbreak sickened 605 people and ultimately killed 4 (Roebuck, 2004).

The cause of the outbreak eventually was determined to be fresh green onions or scallions used in the mild salsa, which was made at individual Chi-Chi's restaurants. To place the magnitude of this outbreak in context, the Centers for Disease Control and Prevention (CDC) estimates that 125,000 to 200,000 cases of hepatitis A occur in the Untied States every year and at least 5% of these cases are related to food-borne transmission ("Attorneys Call," 2003). Outbreaks linked to a single restaurant typically involve 25 to 200 cases on average, noted Llelwyn Grant, spokesman for the CDC (Snowbeck, 2003).

An examination of the chronology of the event illustrates Chi-Chi's communication responses throughout the crisis cycle. Chi-Chi's Chief Operations Officer (COO), Bill Zavertnik, was the organization's primary spokesperson, presiding at the press conferences held by the company. The only other public message came in the form of a brief statement by a divisional vice president at the identification of the crisis (Table 5.1).

TABLE 5.1.
Chronology of Chi-Chi's Hepatitis A Outbreak

November 2, 2003	Beaver City Mall restaurant voluntarily closes.
November 3, 2003	Pennsylvania Department of Public Health issues a hepatitis A advisory to people who might have eaten at the Beaver Valley Mall Chi-Chi's between October and November 2.
November 4, 2003	Steven Sprowl, divisional vice president, emphasizes that outbreak is an isolated incident and that the restaurant has scored 100% on recent health department inspections. He also states that the restaurant would not reopen until the management is confident that the health and safety of their employees and guests are not in question (Snowbeck, 2003).
November 5, 2003	4,000 people showed up for vaccines being administered by the Health Department at a local college.
November 6, 2003	Confirmed hepatitis A cases rise to 130.
November 7, 2003	Chi-Chi's distributes a press release, expressing concerns regarding the recent outbreak of hepatitis A.
November 10, 2003	Chi-Chi's announces it would hold a press conference on November 12.
November 11, 2003	Chi-Chi's postpones news conference until more definitive data become available. Preliminary investigations focus on problems with restaurant employee hygiene. Chi-Chi's announces that, "after consultation with the Centers for Disease Control (CDC) and the Pennsylvania Department of Health, the decision had been made to delay the press conference at which the company planned to address the recent outbreak of hepatitis A. In the last two years, our Beaver Valley location has had four separate health inspections and we have scored 94, 95, 98, and 100 out of a possible 100" (Chi-Chi's.com, 2004).
November 12, 2003	Chi-Chi's announces it will voluntarily remove green onions from every menu item at all locations in the restaurant chain. It explains that it's decision reflects the fact that contaminated green onions are the prime suspect of recent hepatitis A outbreaks in various other states (Chi-Chi's.com, 2004).
November 13, 2003	Chi-Chi's announces it will open its Beaver Valley Mall restaurant for at least 60 days.
November 19, 2003	CCG Strategic Communications is brought in to deal with the victims. The public relations firm sends a staffer to Pennsylvania to coordinate with Chi-Chi's, the CDC, and the Pennsylvania Department of Health (O'Dwyer, 2003).

November 21, 2003	Chi-Chi's holds news conference, stating that a shipment of green onions was the likely source of the hepatitis A outbreak and that nothing could have been done to prevent it. A toll-free number for customers is announced. COO Zavertnik said, "as soon as there was a suspicion of green onions being involved, Chi-Chi's removed them from our entire chain. There is currently no industry-accepted means of testing produce for hepatitis A virus. And there is no effective way to wash hepatitis A off green onions, as the public health authorities have confirmed" (Chi-Chi's, 2004). Chi-Chi's asks a bankruptcy court for permission to pay a $500,000 insurance deductible to free up as much as $51 million in insurance to settle customer and employee claims. The bankruptcy court gives limited approval to Chi-Chi's plans to begin paying some expenses, up to $20,000 per claim, of those sickened in the outbreak.
November 21, 2003	The CDC announces that green onions served at Chi-Chi's were implicated in the deaths of three, 605 illnesses, and 9,100 Immune Globulin shots.
November 22, 2003	Chi-Chi's CEO Bill Zavertnik makes a brief statement and explains that an investigation by the Centers for Disease Control and Prevention and Pennsylvania Department of Health found no wrongdoing by company. Zavertnik takes no questions from reporters.
November 30, 2003	Attorneys from Marler Clark, a Seattle-based law firm specializing in food illness-related lawsuits, files lawsuits against four suppliers of the tainted onions. The previous lawsuits the firm filed against Chi-Chi's are dismissed without prejudice.
December 1, 2003	COO Zavertnik announces that, "the dismissal of the lawsuits confirms our belief that Chi-Chi's could not have done anything to prevent such an outbreak from occurring. However, we are deeply concerned with the impact this incident has had on our guests and employees and we are committed to rebuilding the trust of the community. We hope our guests will take advantage of the outreach effort and use the 1-800 number to address questions and concerns they may have and request our assistance" (Chi-Chi's, 2004).
February 17, 2004	COO Zavertnik says his company has paid out less than $1 million to settle claims related to the hepatitis A outbreak.

TABLE 5.1.
Chronology of Chi-Chi's Hepatitis A Outbreak *(continued)*

February 19, 2004	A bankruptcy judge approves an agreement to create a mediation system to handle hundreds of legal claims for people sickened during the hepatitis A outbreak from November.
April 1, 2004	Fourth victim of hepatitis A outbreak dies (Roebuck, 2004).
July 19, 2004	Bankruptcy court authorizes Chi-Chi's to pay $2 million in hepatitis settlements ("Court OKs Chi-Chi's," 2004).
August 24, 2004	Chi-Chi's announces it is suing food wholesalers to help with lawsuit payments ("Chi-Chi's Sues," 2004).
September 23, 2004	Outback Steakhouse purchases remaining restaurants in Chi-Chi's chain (Schreiner, 2004).

PRECRISIS CONDITIONS

Analyses of the precrisis conditions at any organization typically reveal some level of operational and leadership deficiency. In most cases, such deficiencies mean that managing the crisis is more challenging than necessary, and in others, it can mean there is nothing the organization can do to survive. In the case of Chi-Chi's, three conditions presented significant challenges. First, the company's business structure failed to do everything it could to guarantee food safety. Second, Chi-Chi's and its parent company were experiencing serious financial challenges.Third, the company had not developed adequate relationships with one of their key stakeholders—their suppliers.

Restaurant owners and managers have many options in their precrisis planning. Chain restaurants can be mandated to develop specific plans by their corporate headquarters or they may be given a plan to use. In turn, the corporate operation or headquarters may be owned by a parent company that provides direction to the local units. The approach a given company takes to precrisis planning and the success of the plans are influenced by the company's business model, the onsite restaurant managers, business pressures, public relations philosophy, and generally accepted restaurant practices.

Chi-Chi's business model is outlined by its parent company, Prandium, Inc. The Web site relates that the restaurant operations have one common goal "that unites us—namely, making Enjoyment the Main Course for our guests" (Prandium, 2004). The company values include safety, respect, trust, service, success, and fun. Safety is the relevant value to this discussion. The company specifically states, "they are committed to the safety of our products and facilities for our team members and our guests" (Prandium, 2004). It is fairly typical within the restaurant industry that safety is listed as a first value, often to create and maintain a culture of safety and to reassure both investors and customers.

Listing the safety value on the corporate Web site, however, is not enough to prove that the organization is committed to safety in its practices. *Safety* is a broad term and can apply to a variety of issues. First, it can apply to worker safety in terms of accidents in the restaurant. Second, it can apply to consumer safety, such as the physical safety of the restaurant. Third, it can apply to the raw materials utilized to make the consumer's meal. Finally, the mission statement or company values can be applied to production processes, themselves, including transportation, cooking, and food safety. Inclusion of a term in a corporate mission or value statement does not necessarily translate into consistent employee practices. There are specific questions that can be asked to help evaluate the extent the company mission translates to best practices in safety. Does the definition of safety extend to an enforcement of supplier standards? How is supplier safety performance utilized in awarding future contracts? What level of safety training is provided for new employees? What ongoing training is provided? Does the company have an internal safety audit program? Does the company include safety audit scores as part of its management bonus program? Moreover, the overall company approach to food and employee safety would influence its development of a crisis plan.

Although Chi-Chi's does refer to Department of Health audits (in its early communication about the outbreak), the company does not mention or use an internal audit program as part of its safety program. In light of continuing media coverage of issues such as sanitation, most of the food industry has some program of self-regulation. Such self-regulation, often through audits conducted by the company's quality control department, is intended to bring problems to light before official, external agency audits take place. Chi-Chi's failure to highlight such an audit program leads to speculation that either the audits did not take place or that the scores were low. Under any other scenario, the internal audits would have likely been highlighted in the company's press releases.

Chi-Chi's also did not refer to the strength of their employee training in its communication to the public. Their quality assurance department's work was not used to validate their food storage and handling procedures. It is customary for organizations to use training as one of their first lines of defense when facing attack. For example, if a company is sued for discrimination, its public relations department is likely to act quickly to prove that all of their employees went through training and explaining the company's position against such a behavior. Based on what Chi-Chi's failed to offer the training of their employees as a defense, it may be that the definition of the safety value in the mission does not include food safety.

The management team is another variable impacting a company's crisis plan. Although a corporation may distribute a plan and instruct managers to implement it, ultimately the onsite manager must determine the level of emphasis placed on the plan. If Chi-Chi's restaurant or regional managers are typical, they have not received adequate training in leading the employees in their operation, let alone managing a crisis situation. The restaurant industry, which often boasts of detailed and effective training for hourly workers and unit managers, falls short when it comes to teaching leadership skills to middle- and senior-level managers, according to a study by the Elliot Leadership Institute at Johnson & Wales University (Berta, 2003). Leaders are being trained to manage, not to lead. This emphasis on management focuses on daily operational tasks such as scheduling and meeting budget expectations, but may ignore the higher level skills demanded by a crisis.

A third factor contributing to precrisis planning is the overall business condition being faced by the organization. Businesses facing serious financial challenges, as Prandium was, understandably make keeping the company solvent the top operational priority. Given that their continued existence was in question, Chi-Chi's communication with stakeholders likely spent more time answering questions about the future, rather than detailing quality assurance efforts or safety contingency plans.

The fourth factor influencing the effectiveness of precrisis planning is the corporation's philosophy of public relations. On a routine basis, corporate public relations departments handle questions about the company's product or service and look for opportunities to promote the corporate image. Facing a crisis presents a strong test, much different than routine daily communication. An even bigger test may have been encountered by the public relation manager's success in establishing and maintaining a crisis plan. In addition, the philosophy on calling in external public relations help is a variable and is debated within most industries. Prandium chose the approach of bringing in a company, CCG Strategic Communications, a Sherman Oaks, California, public relations company they had worked with previously (Jack O'Dwyer's Newsletter, 2003). David Watson, director of CCG's corporate branding practice, was hired after the CDC originally reported that 127 residents in the Beaver Area may have fallen ill with hepatitis A. Another implication of hiring an external firm during a crisis is that the organization may be trying to separate them from the problem. The corporate explanation regarding that issue is that their internal resources need to focus on resolving the actual problem. Chi-Chi's offered such an explanation by noting on November 22, 2003, that over the past 2 weeks, "we at

Chi-Chi's have spent every waking hour working with public health" (Zavertnik, 2003).

The final variable affecting precrisis planning may be the most significant. Businesses put most of their efforts into the factors they think may put them at the greatest risk. The food industry focuses most of its efforts on the issues thought to be the primary causes of food-related illness. The two factors that are most commonly reported as contributing to food-borne illness are improper holding and storage temperatures and poor personal hygiene among food handlers (Collins, 2001).

Another standard of excellence in the food industry is to hold suppliers of fresh and processed products to high standards. This is often accomplished through supplier visits and inspections, and determining whether if the supplier follows generally accepted industry standards. The restaurant may also ask for records of the supplier's compliance with processing standards.

In Chi-Chi's case, there was a challenge in ensuring their supplier's attention to standards. Chi-Chi's obtained green onions through a food broker, who imported the product from Mexico (Hodge, 2003). There is no record of Chi-Chi's monitoring the standards of the Mexican suppliers or their broker.

The other generally accepted business practice is using the heat of cooking the food product to provide the ultimate check against most food-borne illness. Products that are served raw are generally a small part of the menu and food costs, so they do not receive as much attention as, for example, meat or fish products. Although Chi-Chi's business practices may not be unique within the casual dining industry, there is no evidence that they promoted or practiced any consistent monitoring of their fresh products.

In its early days, Chi-Chi's achieved some financial success. In 1972, Chi-Chi's began as an independent casual dining Mexican restaurant. The company was relatively unique, providing inexpensive Mexican cuisine and an entertaining dining experience. Due to its great success, the restaurant was immediately franchised and grew to 92 restaurants by 1982, with revenues of $100 million. By the mid-1980s, however, the company's financial performance turned downward. In November 1984, the stock was trading for $10 per share, less than half the price of January 1984.

In attempts to rescue the chain, a long list of parent companies became involved in Chi-Chi's operations. Kelly-Johnston Enterprises was the first, filing Chapter 11 bankruptcy in 1985 (Kelly-Johnston, 1985). Foodmaker, Inc. became the owner in 1988 and sales dropped 55%. In 1993, two thirds of the Chi-Chi's chain was sold, and Foodmaker merged with the Restaurant Enterprise Group to form Family Restaurant's, Inc. Two years later, Family merged with Koo Koo Roo, a regional Southern California operation known for its rotisserie-cooked and steam-roasted poultry products and changed the corporate name to Prandium in 1998 (*Courier-Journal*,

2004).

Although many efforts were made to change Chi-Chi's and improve the chain's revenue picture, Prandium entered Chapter 11 bankruptcy proceedings in May 2002. All three of Prandium's subsidiaries have filed for bankruptcy protection, including Chi-Chi's, who entered Chapter 11 bankruptcy proceedings in October 2003. The constraints imposed by bankruptcy became significant in Chi-Chi's ability to respond to the financial challenges presented by the hepatitis A outbreak.

Chi-Chi's financial picture is a significant precrisis condition in two ways. First, the string of parent companies undoubtedly caused confusion within the organization. Every time a new parent company comes in, they make their mark through personnel and policy changes. At a minimum, these changes prohibited Chi-Chi's from creating a consistent strategy and returning to profitability.

Second, cost-cutting measures are an inherent element of any financial recovery. Increasing revenues and reducing expenses are the two ways to improve the financial performance of any business. Seeger et al. (2003) reinforce the connection between financial struggles and crisis planning by observing, "when faced with budget constraints, for example, risk mitigation strategies and structures are often seen as unnecessary luxuries and are reduced or eliminated" (p. 107). Although there is not specific evidence to demonstrate that crisis planning or risk management was a victim of financial cuts, staffing cuts are often employed as a quick way to reduce expenses.

Regardless of whether or not a company is facing financial problems, suppliers are an important part of any business. They are particularly important in the food business. Without suppliers, no food can be prepared, no meals can be served, and no revenue can be earned. Business owners and managers look to their suppliers to give them the best price possible, and trust that they work as partners, such that the suppliers' profits grow in the same pattern as the restaurant sales grow.

Supplier relationships need to be built and nurtured on a regular basis, because organizations do not know when extraordinary conditions will demand more of the relationship. Although crises tend to appear randomly, their presence should not be unexpected (Hearit & Courthright, 2003). If Chi-Chi's had supportive and cooperative supplier relationships, they should have been aware of the problems that other regional restaurants were having with green onions.

A preliminary investigation by the FDA showed that the virus infecting the Chi-Chi's customers is closely related to one isolated from victims of hepatitis A outbreaks caused by scallions in Tennessee and Georgia in September. In those cases, the scallions were traced to suppliers in Mexico (Grady, 2003). In Knoxville, Tennessee, 80 people became sick with hepatitis A in September after eating at one city restaurant, said Mark Jones, director of the Knox Country Health Department. Contaminated green onions are the prime suspects in both states (Snowbeck, 2003). O'Charleys is a 203-unit chain operation in 16 states and was the suspected source of the Tennessee outbreak in September (Allen et al., 2003).

In Georgia, about 250 people came down with hepatitis A in September and October after eating at about 12 restaurants in the central and northern parts of the state, said Richard Quartarone, spokesman for the state's health department. Another outbreak, which sickened 16 people in North Carolina, probably stemmed from the same source as the Georgia outbreak, according to a report by Georgia's health department (Mandak, 2003).

Given this readily available information, it is somewhat surprising that Chi-Chi's suppliers did not inform the company of these problems. By not reporting problems in other restaurants using raw green onions, the supplier put Chi-Chi's in a vulnerable position. The lack of strong relationships between Chi-Chi's and its suppliers became an insurmountable precrisis condition.

POST-CRISIS EFFORTS

The precrisis conditions of bankruptcy, poor supplier relations, and inadequate food safety procedures were so overwhelming that Chi-Chi's was significantly limited in its ability to respond to the crisis. An examination of their efforts at salvaging legitimacy and creating organization illustrate's these limitations.

An organization may attempt to salvage legitimacy by showing that it acted without being forced to, whether it is a product recall, a disaster evacuation, or a restaurant closing. The question of the timing of Chi-Chi's decision to close the Pennsylvania restaurant is worth closer examination. The company communication consistently points to the voluntary closing, perhaps hoping for a public response of understanding and respect for the decision. However, the Chi–Chi's closing followed the Pennsylvania Department of Health notification that the restaurant was the probable source of infection (Polgreen, 2003). Given the amount of investigation, it is likely there were signs pointing to the restaurant prior to the date of the official Department of Health statement.

Chi-Chi's executives made several public comments regarding their concern for employees, the consumers who had eaten at their restaurant, and their larger community. However, little attention was devoted to their larger social responsibility. Most of the content emphasized that the organization had done nothing wrong. The organization had many opportunities to say that they wanted to resolve the issue, regardless of the cause. In addition, when the cause was traced to the imported green onions, apparently contaminated prior to the time Chi-Chi's received them, the corporate messages began reinforcing that nothing can be done to test produce for hepatitis.

Another strategy to salvage legitimacy could have involved references to the quality of the suppliers used to provide Chi-Chi's with green onions. Throughout the investigation, the organization made no comments about trusting their suppliers or the process by which they ensure supplier quality. This particular stakeholder group was left out of all communication until they were identified as the likely source of the illness.

Given that the ultimate cause, of hepatitis A can most likely be traced to the growing conditions, a site visit might have determined that the supplier was not following recommended procedures. Many restaurants design their food safety practices so corporate spokespersons or advertising can argue that they go beyond required standards. Chi-Chi's did make these kinds of arguments. Within the food industry, it is common to have some kind of supplier rating system, with many larger chains performing regularly scheduled supplier site visits. Some large restaurant chains have the clout to demand a detailed accounting of the manufacturing process related to ready-to-eat products they buy from distributors or processors. Others have the resources to inspect production facilities and supplying farms first hand, according to Steven Grover, the National Restaurant Association's vice president for health and safety ("California Operators," 2003). According to Devon Zagory, senior vice president of the food safety firm Davis Fresh Technologies LLC of Redding, California, the industry standard for large grocery and restaurant chains is to audit their suppliers' compliance with the FDA guidelines (Mandak, 2003).

Due to Chi-Chi's financial situation and its failure to build or optimize supplier relations, the company was in a difficult position to salvage its legitimacy. Quick settlements could not be made with victims of the outbreak, which allowed more time for lawsuits to develop. Chi-Chi's could not team with one of their prime stakeholder groups, suppliers, in demonstrating its innocence in causing the food-borne illness.

MISSED OPPORTUNITIES

The Chi-Chi's spokespersons faced several unique communication challenges that could have been turned into opportunities. First, the law firm representing a majority of the victims, Marler Clark, took an aggressive public communication stance. The focus of Marler Clark's practice is representing victims of food-borne illness; the two partners were opposing counsel in the 1993 Jack-in-the-Box e.coli case. Members of their firm wrote press releases and documented the pain and suffering of the victims, along with Chi-Chi's financial problems. These media pieces were then posted on the company's Web site. Chi-Chi's could have used their Web site to further their cause, but failed to institute an aggressive proactive strategy.

Chi-Chi's also could have further affirmed its social legitimacy by making some public commitment to demonstrate their determination to keep a similar outbreak from occurring again. For example, Marler Clark, the law firm representing approximately 80 of the outbreak's victims, is pushing for states to require hepatitis A vaccines for all food handlers. In fact, this regulation is already in place for culinary staffs in Nevada ("Boskovich Among," 2003). Chi-Chi's never made any comments about improving food sanitation, processing, testing, or importation procedures.

Another way an organization might demonstrate social responsibility is through quick financial reimbursement to victims who lost time from work or by donations to local community organizations. Due to their status in bankruptcy proceedings, Chi-Chi's could not readily volunteer financial payments to victims. Although Chi-Chi's did make some statements of sympathy and concern, they took no actions to demonstrate the sincerity of their verbal comments. Their efforts at affirming their social responsibility were limited and, therefore, of limited effectiveness.

Although no organization would actively seek out a crisis, the aftermath of a crisis does present an opportunity for organization learning. An example of this can be seen in Jack-in-the-Box's actions in the aftermath of its 1993 e. coli outbreak. A week after the outbreak, a food safety expert began working with the chain and later became the chain's director of quality insurance. He changed meat suppliers immediately. The San Diego-based burger chain became a crusader for higher food service standards across the entire food service industry (Bertagnoli, 1996).

Chi-Chi's had the opportunity to instigate similar changes in the produce industry. Although the CEO was quoted as saying there is no industry-accepted method of testing and the restaurant scored well on its inspections, others have found that there was an opportunity for improvement in restaurant operations. Several characteristics of the way food was prepared and served could have contributed to the outbreak's size, including (a) multiple opportunities for intermingling of uncontaminated and contaminated green onions in a common bucket for 5 days with the ice in which they were shipped, and (b) serving contaminated items with a relatively long shelf life (e.g., mild salsa) to a large proportion of patrons over several days (Dato, 2003).

In examining the Chi-Chi's case, health experts have now argued that food preparation at the restaurant amplified the problem. Bunches of green onions were stored together in large buckets for 5 days or more with the ice they had been shipped in from Mexico. As a result, even if only some bunches were tainted with the hepatitis virus when they were delivered, it would have quickly spread to all the other scallions (Grady, 2003).

Chi-Chi's also could have taken a position on increasing testing of imported produce, as less than 2% of the produce that crosses into the United States in inspected for disease-causing bacteria. Current shipments of green onions are being closely scrutinized (Lindquist et al., 2003). Given that more work is needed in terms of identifying viral contamination in produce, Chi-Chi's had the opportunity to lead those efforts and has thus far failed to do so.

Donna Garren, vice president of scientific and technical affairs for the United Fresh Fruit and Vegetable Association, provides another example of the learning that could have taken place at Chi-Chi's. Garren notes, when the dust settles from the hepatitis A outbreak, buyers often will need to start redefining expectations in their contracts. Those expectations will have to be back to their primary supplier, telling the suppliers they need to eliminate players, be it a harvest crew, farm, and so on who don't achieve some kind of standardization in terms of sanitation hygiene (McTaggart, 2004).

The hepatitis A outbreak in Monaco, Pennsylvania, Chi Chi's restaurants is not the first crisis affecting Philadelphia and its surrounding area. Outpouring of local support for the restaurant was undeniable. The area's past experience with organizational crises might provide part of an explanation. Chamber of Commerce President Gitnick says area businesses have rebounded, having learned survival skills from steel mills closing and U.S. Airways cutbacks at Pittsburgh International Airport ("Chi-Chi's Closes," 2004). Some lawyers in the area say the hepatitis A outbreak is the biggest source of public health litigation in a generation, ever since the steel industry began generating workers' asbestos claims ("Pa. Town," 2004).

By the time Chi-Chi's reopened the Beaver Valley Mall restaurant, local sentiment toward the chain restaurant appeared to come full circle—from anger and antipathy to forgiveness and acceptance ("Pa. town," 2004). People were waiting in line to get into Chi-Chi's when it reopened. However, after the initial interest, the organization's COO Zavertnik said business has continued to lag since the restaurant opening in mid-January ("Chi-Chi's Closes," 2004).

During the immediate time following the hepatitis A outbreak, there was a drastic decrease in business sales. The national hepatitis A publicity also caused a significant reduction in sales revenue at virtually all Chi-Chi's locations nationwide ("Disease Waning," 2003). Following the crisis, the restaurant chain has closed 26 underperforming restaurants, including 7 in Pennsylvania. The Beaver Valley restaurant was not among those being closed (Chi-Chi's, 2003). These closings led to industry speculation about the future of the restaurant chain.

There is no longer any reason to speculate about Chi-Chi's long-term viability as a restaurant chain. On September 23, 2004, Outback Steakhouse,

Inc. announced that it had "closed the deal for the rights to 76 restaurants in the Chi-Chi's chain" (Schreiner, 2004). Outback's purchase did not include Chi-Chi's brand, and the company plans to use the locations for their own brands (Schreiner, 2004). Both the company's Web site and customer service number expressed thanks to their customers. If Chi-Chi's had not closed at this time, the dollar value of the lawsuits following the incident probably would have forced them to do so eventually. The company's attorneys said they have paid more than $40 million in lawsuits resulting from the outbreak (Associated Press, 2005).

Many factors impact an organization's ability to successfully respond to the inevitable crisis. When contrasted with other restaurants that have faced significant food-borne illness outbreaks, such as Jack-in-the-Box and Schwann's, Chi-Chi's precrisis conditions were too much for the organization to overcome. The pre-crisis conditions of a bleak financial picture and inadequate supplier relationships may not have been faced by other restaurants and offer the most significant explanation for Chi-Chi's crisis management problems. This suggests that precrisis conditions cannot always be overcome through communication efforts.

IMPLICATIONS

There are several implications for the food service industry and communication research. Trust is critically important in the food industry. The belief that a food product or provider is unsafe or untrustworthy can undermine the fundamental willingness of customer's to buy those products. The threat of lost legitimacy and economic disruption is probably as great as the threat of actual harm arising from a food issue. Being associated with a food-borne illness is one of the most serious challenges a restaurant can encounter. The potential loss of human life, the financial exposure due to lawsuits, and the threat to the future viability of the restaurant are real issues facing the individual restaurant, its employees, and owners. To avoid such negative consequences, restaurants must closely review their supplier relationships and utilize those suppliers in ensuring a safe food supply chain.

In addition, as the demand for fresh food products continues to grow, more attention needs to be paid to inspection and food-handling procedures for produce. The Jack-in-the-Box e. coli outbreak highlighted the need for proper cooking temperatures, and the Chi-Chi's outbreak should do the same for produce handling. Communication and business scholars should continue to research the significance of precrisis stakeholder relationships. The value of employee and customer relations is well accepted, but the area of supplier relationships offers potential for future learning.

REFERENCES

Allen, R. L., Gould, A., Koteff, E., & Martin, R. (2003). Rush to judgment in aftermath of health flaps stings operators, opinion shapers. *Nation's Restaurant News*, *37*, 27.

Associated Press. (2005, September 30). *Hepatitis victim settles Chi-Chi's suit.* Retrieved November 1, 2005, from http://marlerclark.com/news/chichis104.htm

Attorneys call for mandatory hepatitis A vaccinations for all foodservice workers. (2003, November 4). Retrieved April 15, 2004, from http://marlerclark.com.

Benoit, W. L. (1995). Sears' repair of its auto service image: Image restoration discourse in the corporate sector. *Communication Studies, 46*(1&2), 89–105.

Benson, J. A. (1988). Crisis revisited: An analysis of strategies used by Tylenol in the second tampering episode. *Central States Speech Journal, 39*(1), 49–66.

Berta, D. (2003, May 26). Institute's study find midlevel managers lack leadership training. *Nation's Restaurant News*. Retrieved May 8, 2004, from http://www.findarticles.com/cf_dls/m3190/21_37/102659626/print.html.

Bertagnoli, L. (1996). After the outbreak: Jack in the Box's quick response to disaster has turned into a crusade for food safety, *Restaurants & Institutions*, *106*, 92.

Boodman, S. (2003, December 2). Focusing on the safety, of imported produce. *Washington Post*, p. A47.

Boskovich among distributors named in lawsuit over hepatitis. (2003, December 2). *Ventura County Star*. Retrieved April 15, 2004, from http:://80-web.lexis-nexis.com/ezproxy.emich.edu.

Brinson, S. L., & Benoit, W. L. (1999). The tarnished star: Restoring Texaco's damaged public image. *Management Communication Quarterly, 12*, 483–510.

California operators wary of pre-washed produce's cleanliness. (2003, December 8). *Nation's Restaurant News, 37*(49), 57.

Chi-Chi's. (2003). *Company information*. Retrieved March 28, 2004, from http://www.chi-chi's.com.

Chi-Chi's closes 26 ailing restaurants. (2004, February 28). *The Patriot*. Retrieved March 26, 2004, from http://marlerclark.com.

Chi-Chi's pays out less than $1 million. (2004, February 16). *Charleston Daily Mail*. Retrieved March 26, 2004, from http://marlerclark.com.

Chi-Chi's sues suppliers in green onions hepatitis A outbreak. (2004, August 24). *T&B Outbreaks Week*, p. 22.

Collins, D. J. (2001). Impact of changing consumer lifestyles on the emergence/reemergence of foodborne pathogens. *Emerging Infectious Diseases, 3*(4), 1–13.

Coombs, W. T. (1995). Choosing the right words: The development of guidelines for the selection of the appropriate crisis-response strategies. *Management Communication Quarterly, 8*(4), 447-476.

Courier-Journal. (2004). Retrieved February 19, 2004, from http://www.courier-journal.com/business/news2003/11/30/bis-front-chichi30-6787.html.

Court OKs Chi-Chi's $2 million in hepatitis settlements. (2004, July 19). *Health & Medicine Week*, p. 630.

Dato, V., Weltman, A., Waller, K, Ruta, M. A., Hembree, C., Everson, S., Wheelter, C .& Vogt, T. (2003, November 28). Hepatitis A outbreak associated with green onions at a restaurant. *Morbidity and Mortality Weekly Report, 522*, 1155.

Disease waning, but costs continue to climb. (2003, November 27). *Pittsburgh Post-Gazette*. Retrieved March 26, 2004, from http://marlerclark.com.

FDA, Mexico disagree on source of hepatitis. (2003, December 10). *Charleston Gazette*. Retrieved March 26, 2004, from http://marlerclark.com.

Grady, D. (2003, November 22). Government makes it official: Blame scallions for outbreak. *New York Times*, to A-13. Retrieved April 9, 2004, from http://80-web.lexis-nexis.com.

Hearit, K. M. (1997). On the use of transcendence as an apologia strategy: The case of Johnson Controls and its fetal protection policy. *Public Relations Review, 23.*

Hearit, K. M., & Courthright, J. L. (2003). A social constructionist approach to crisis management: Allegations of sudden acceleration in the Audi 5000. *Communication Studies, 54*, 79.

Industry works to reform image of food safety. (2003, September 22). *Nation's Restaurant News, 37*, 122.

Johnson, C. E., Sellnow, T. L., Seeger, M. W., Barrett, M. S., & Hasbargen, K. C. (2004). Blowing the whistle on Fen-Phen: An exploration of MeritCare's reporting of linkages between Fen-Phen and valvular heart disease. *Journal of Business Communication, 41*(4), 350-369.

Kelly-Johnston board may face challenges from 2 stakeholders. (1985, May 13). *The Wall Street Journal* (Eastern edition), p. A1.

Lindquist, D., Dibble, S., & Walker, S. L. (2003, December 1). A bumper crop of fear, anxiety: Investigators go to onion fields on trail of deadly hepatitis outbreak. *San Diego Union-Tribune*. Retrieved April 5, 2004, from http://marlerclark. com.

Mandak, J. (2003a, November 22). *Chi-chi's executive calls restaurants safe.* Retrieved April 15, 2004, from http://marlerclark.com.

Mandak, J. (2003b, November 22). *Chi-Chi's litigation hinges on company's insurance, emergence from Chapter 11.* Retrieved April 10, 2004, from http://80-web.lexis-nexis.com.

Mandak, J. (2003c, November 23). *Chi-chi's litigation hinges on company's insurance.* Retrieved April 15, 2004, from http://marlerclark.com.

Manning, A., & Weise, E. (2003, November 19). Hepatitis A outbreak tied to imported food. *USA Today*. Retrieved April 12, 2004, from http://marlerclark.com.

Martin, R. (2003). Chain pulls scallions amid epidemic. *Nation's Restaurant News, 37*(48), 1.

McTaggart, J. (2004, January 1). Produce's paradigm shift: An up tick in food borne illness puts the produce supply chain in the spotlight. *Progressive Grocer, 83*, 69.

Millar, D. (2003). *ICM crisis report: News coverage of business crises during 2002.* Retrieved May 7, 2004, from http://www.crisisexperts.com.

O'Dwyer, J. (2003). Chi-Chi's hires CCG for crisis. *Jack O'Dwyer's Newsletter, 35*(45), 2.

Pa. Town, restaurant trying to bounce back, Official says people no longer blame chain for outbreak. (2004, January 31). *Charleston Daily Mail*, p. 3c.

Pauchant, T. C., & Mitroff, I. I. (1992). *Transforming the crisis-prone organization.* San Francisco: Jossey-Bass.

Pittsburgh Post-Gazette. (2004, February 15). Retrieved April 9, 2004, from http://80-web.lexis-nexis.com.

Polgreen, L. (2003, November 17). Community is reeling from hepatitis outbreak. *New York Times*. Retrieved April 10, 2004, from http://marlerclark.com.

Prandium. (2004). *Prandium, Inc. Chi-Chi's company information*. Retrieved April 8, 2004, from http://www.prandium.com.

Ray, S. J. (1999). *Strategic communication in crisis management: Lessons from the airline industry.* Westport, CT: Quorum Books.

Roebuck, K. (2004, April 6). Hepatitis A menace still looms months later. *Pittsburgh Tribune-Review*. Retrieved April 24, 2004, from http://marlerclark.com.

Schreiner, R. (2004, September 23). Outback finalizes Chi-Chi's purchase. *Cincinnati Post*. Retrieved September 30, 2004.

Seeger, M. W. (1986). The Challenger tragedy and search for legitimacy. *Central States Speech Journal, 37*, 147–157.

Seeger, M. W., Sellnow, T., & Ulmer, R. (2003). *Communication and organizational crisis*. Westport, CT: Praeger.

Seeger, M. W., & Ulmer, R. R. (2001). Virtuous responses to organizational crisis: Aaron Feuerstein and Milt Cole. *Journal of Business Ethics, 31*, 369-376.

Seeger, M.W., & Ulmer, R. R. (2003). Explaining Enron: Communication and responsible leadership. *Management Communication Quarterly, 17*(1), 58–85.

Sellnow, T. L., & Brand, J. D. (2001). Establishing the structure of reality for an industry: Model and anti model arguments as advocacy in Nike's crisis communication. *Journal of Applied Communication Research, 31*, 278–295.

Sellnow, T. L., Ulmer, R. R., & Snider, M. (1998). The compatibility of corrective action in organizational crisis communication. *Communication Quarterly, 46*, 60–74.

Snowbeck, C. (2003, November 7). *Hepatitis outbreak in Beaver Country reaches 130: One may need liver transplant; 2 or 3 others in intensive care*. Retrieved April 10, 2004, from http://marlerclark.com.

Snowbeck, C. (2003, November 13). *Beaver County hepatitis probe changes focus*. Retrieved April 10, 2004, from http://marlerclark.com.

Turner, B. (1976). The organizational and interorganizational development of disasters. *Administrative Science Quarterly, 21*, 378–397.

Ulmer, R. R.(2001). Effective crisis management through established stakeholder relationships: Malden Mills as a case study. *Management Communication Quarterly, 14*(4), 590–615.

Ulmer, R. R., & Sellnow, T. L. (2000). Consistent questions of ambiguity in organizational crisis communication: Jack in the Box as a case study. *Journal of Business Ethics, 25*(2), 143–155.

Weick, K. E. (1993). The collapse of sensemaking in organizations: The Mann Gulch disaster. *Administrative Science Quarterly, 38*, 628–652.

Williams, D. E., & Treadaway, G. (1992). Exxon and the Valdez accident: A failure in crisis communication. *Communication Studies, 43*, 56–64.

Zavertnik, B. (2003). Retrieved April 1, 2004, from http://www.chi-chis.com.

PART IV

Infectious Disease Outbreaks

6

BEST PRACTICES IN PUBLIC HEALTH COMMUNICATION

Managing West Nile Virus in Arkansas From 2002-2003

Robert R. Ulmer

Robert J. Alvey

Jennifer Kordsmeier

This chapter examines the Arkansas Department of Health's response to the 2002-2003 West Nile Virus (WNv) outbreak. We use best practices in risk and crisis communication to better understand appropriate and effective responses to public health outbreaks. We conclude that the Arkansas Department of Health appropriately responded to the WNv public health event through application of best practices in risk and crisis communication.

Public health officials must communicate publicly about complex, uncertain, unpredictable, and often unprecedented health risks. The purpose of a public health response to a crisis is to efficiently and effectively reduce and prevent illness, injury, and death and return individuals and communities back to normal (Reynolds, 2002). As a result, this communication challenge can be daunting. Public health officials often have limited information or experience with the particular health issue, face an immediate demand for explanations, and yet cannot communicate with any certainty. Moreover, the events are often surrounded by high levels of public concern and, in some cases, anger. Examples such as Mad Cow Disease, SARS, and anthrax exemplify this challenge to public health communication (Covello, 2003; Covello, Peters, Wojtecki, & Hyde, 2001; Sellnow, Seeger, & Ulmer, 2005; Vanderford, 2003).

This chapter examines the Arkansas Department of Health's (ADH) response to the 2002-2003 WNv outbreak in Arkansas through a discussion of seven best practices in risk and crisis communication. The goal of this chapter is to examine the realities of public health communication during a challenging and complex crisis. In doing so, we provide a discussion of several key risk and crisis communication principles and best practices. Next, we provide a brief context of WNv in Arkansas. Third, we examine ADH's preparation and response to WNv. Finally, we offer recommendations and implications for communicating about public health crises.

PUBLIC HEALTH RISK/CRISIS COMMUNICATION

As the fields of risk and crisis communication become more mature, there is considerable discussion about whether an event is labeled *risk communication* or *crisis communication* (Reynolds & Seeger, 2005; Seeger, Sellnow, & Ulmer, 2003). Many public health organizations are uncomfortable with the connotation associated with a crisis. Crisis may create the impression that the public health risk is somehow out of control. Public health departments, to connote a more controlled and less severe event, often use the term risk. Crisis communication scholars also wrestle with the issue of differentiating between risk and crisis communication. Risk communication is often discussed as "the exchange of information among interested parties about the nature, magnitude, significance or control of a risk" (Covello, 1992, p. 359). *Crisis communication* is defined as to "protect health, safety, and the environment by keeping the public informed" and "to restore public confidence in the organization's ability to manage an incident" (Coombs, 1999a, 1999, p. 4). In this chapter, risk communication takes place in the precrisis stage of a public health event, and crisis communication takes place after the trigger event. This view is consistent with the CDC's conceptualization of crisis and emergency risk communication.

Research on risk and crisis communication suggests that there are some general principles that are accepted communication standards for both risk and crisis communication (Covello, 2003). Covello describes seven best practices for risk and crisis public health communication:

1. Accept and involve people or organizations that have a special connection to your organization.
2. Listen to people.
3. Be truthful, honest, frank, and open.
4. Coordinate, collaborate, and partner with other credible sources.
5. Meet the needs of the media.
6. Communicate early and with compassion.
7. Plan thoroughly and carefully.

We examine these seven characteristics of best practices in our discussion of WNv in Arkansas. In doing so, we reorder some of these categories for the sake of clarifying our discussion. We begin with the principle of planning thoroughly. Next we discuss communicating early and with compassion. Within that principle, we examine the characteristics of meeting the needs of the media, of being truthful and honest with people, and listening to the public. Third, we discuss coordinating and collaborating with credible sources. Finally, we examined how the ADH accepted stakeholders as legitimate partners in their communication regarding WNv. What follows is a timeline of the events that led up to the ADH's communication regarding WNv and a discussion of how the data were collected for this study.

WNV IN ARKANSAS

Public health departments have a wide-ranging group of issues to communicate with the public, including SARS, WNv, AIDS, Monkeypox, the Avian Flu, as well as other public health issues. In June 2002, the ADH became aware of the possibility of an outbreak of WNv. WNv "is a mosquito-borne flavivirus and a human, equine and, as recently documented, avian neuropathogen that is maintained in nature by a transmission cycle involving avian reservoirs and primarily Culex mosquitoes" (Fleischauer et al., 2003, p. 94). Before its discovery in the state of New York in 1999, few American public health officials had much experience with this relatively new and exotically named disease. This is important for a couple of reasons. First, the virus was new enough that public health officials were only recently aware that birds could carry the virus. In addition, because WNv is typically endemic to the northern United States, the ADH, although aware of the means of transmission, was neither aware of or prepared for the aggressive spread of the virus through the southern United States and into Arkansas. Second, an exotically named and potentially deadly virus was uniquely positioned to attract a great deal of media attention.

As a result of the potential infection of WNv in Arkansas, ADH immediately assembled a special team of experts set up as an Emergency

Operations Center (EOC)[1] "dedicated to human West Nile surveillance" (Fleischauer et al., 2003, p. 94). The team consisted of a wide-ranging group of individuals, including public health information communicators (PIOs), "the divisions of communicable disease and epidemiology, the state laboratory, the press office, the information technology office, the administration office, and a field assistance team from the United States Centers for Disease Control and Prevention (CDC)" (Fleischauer et al., 2003). This group was assembled to collect information about WNv in Arkansas and communicate to the public its spread and how to take precautions in avoiding contracting the virus. In terms of collecting evidence of WNv in Arkansas, the ADH contracted with the state's Livestock and Poultry Commission (LPC) to test birds, mosquitoes, and horses. In 2002, "as of September 20, the Livestock and Poultry Commission laboratory reported 336 positive birds. . . . Positive birds were found in 48 of the 75 counties in Arkansas. Crows represented 22 percent of the positives, and 78 percent were Blue Jays" (Boozman, 2002). In addition, "Mosquitoes were trapped at 34 different sites. Positive mosquitoes were found at five different locations around the state" (Boozman, 2002). Finally, ADH found that: During 2002, as of September 20, there were at least 130 horses tested for WNv and 56 positive in 23 counties. The fatality rate was 39 percent, with 22 horses having died. . . . Surveillance of horses for Eastern Equine Encephalitis (EEE) has shown 20 cases in seven counties, with 19 of 20 cases being fatal, a 95 percent fatality rate. . . . EEE is more of a threat to humans than WNv since the death rate in infected humans ranges from 30 to 70 percent (Boozman, 2002).

Clearly, WNv was present in Arkansas, and the ADH needed to prepare to communicate to the public about how to lower the risks of contracting the virus.

In August 2002, Governor Mike Huckabee declared "a statewide emergency and provided $1 million in emergency state aid to help counties fight the mosquito-borne disease" (Yee, 2002). The money was used statewide for mosquito abatement. Through this action, the Governor provided some relief to Arkansas residents and the ADH in terms of abating the mosquito population, and hence the opportunity for further spread of the virus. Forty three human cases were discovered (33 West Nile Meningoencephalitis and 10 West Nile Fever) and five deaths were confirmed (Fleischauer et al., 2003). Throughout the event, the ADH provided information to stakeholders and partners (both internal and external) concerning the risks associated with WNv. What follows is a brief description of how the data were collected for this study.

[1]The ADH special team functioned much like an EOC with the exception that it assembled twice a day initially rather than 24/7 and eventually evolved into a daily briefing.

DATA COLLECTION

This study employs a case study method to better understand the ADH's communication during the 2002–2003 WNv outbreak. The ADH provided the funding for the data collection associated with this study. The funding was used for both collection of media accounts and the focused interviews. Interviewees were asked questions regarding their preparation, communication strategies, and organizational efforts for managing the WNv outbreak in Arkansas. Media accounts were used to corroborate and cross-reference the interviews and provide additional detail.

Responding to WNv in Arkansas From 2002 to 2003

Although ADH had an EOC established and the ADH public communication specialists in place, they began a communication awareness effort through the media[2] to make citizens aware of what precautions to take as well as not to overalarm about the risk of contracting WNv. In analyzing this plan, we examine four broad categories of best practices in public health risk and crisis communication. These categories include planning thoroughly; and communicating clearly and with compassion, which includes meeting the needs of the media, being truthful, honest, and open, as well as listening to people. Coordinating, collaborating, and partnering with other credible sources is the third best practice. The fourth, and final goal for effective risk and crisis communication is accepting people and organizations as legitimate partners. What follows is a discussion of the ADH's communication awareness program and an evaluation of that effort through the extant best practices in risk and crisis communication.

Plan Thoroughly

Planning is the quintessential precrisis activity necessary to ensure an appropriate response (Coombs, 1999a, 1999b; Fink, 1986; Pauchant & Mitroff, 1992; Penrose, 2000; Seeger, Sellnow, & Ulmer, 2003). Covello (2003) explains that planning thoroughly involves training staff in "basic, intermediate, and advanced risk and crisis communication skills" (p. 7). In part because of other risks such as anthrax, the ADH had done a great deal of preparation in terms of risk and crisis communication prior to WNv. The two ADH-assigned Public Information Communicators (PIO) assigned to the WNv team were CDC-certified Crisis and Emergency Risk

[2]During the first year of WNv, no funding was available for a campaign that included either public service announcements or awareness material such as posters and T-shirts.

Communications (CERC) spokespersons, and one was a CDC certified CERC trainer. Both PIOs had conducted CERC training for many of the agency's spokespersons and technical staff.

Although ADH did not have a specific plan for WNv, it did have crisis plans in place for disasters at Arkansas Nuclear One in Russellville and the U.S. Army's Pine Bluff Arsenal and began to formulate a specific plan for WNv as soon as the Arkansas risk was apparent. Lee Clark, performance measures leader at the ADH and the WNv special team leader, explained why planning was so difficult: "WNv was so new and different than anything else that we have dealt with. We were taking bits and pieces of emergency response plans and tailoring them. We had to make it into a long-term plan. Our goal was to educate the public and health-care professionals" (personal communication, April 7, 2004). As a result, the ADH's approach to planning was comprised of training communication specialists in risk and crisis communication and diffusing this information to other people in the organization. In addition, ADH used information and planning approaches from established plans at the Arkansas Nuclear One power plant and at a local U.S. Army arsenal dedicated to disposing and treating of chemical weapons. Planning then consisted of the following three strategies.

- ADH provided CDC Crisis and Emergency Risk Communications (CERC) training for numerous agency personnel.
- ADH was involved in continuing education/diffusion of information to co-workers as well.
- ADH adapted existing plans to meet the needs of the WNv event.

These three strategies illustrate a high level of preparation and response capability by ADH. The professional communicators at ADH were training in risk and crisis communication strategies and approaches and were in the process of training other health organizations in the state in these principles. In addition, ADH already had crisis plans in place and were ready to adapt these plans to other potential risks and crises.

Communicate Clearly and With Compassion

A fundamental tenet of crisis communication is to communicate with stakeholders as clearly and with as much compassion as possible (Coombs, 1999a, 1999b; Ulmer, 2001). Covello (2003) explains that risk and crisis communication should use nontechnical language, be clear, and use graphics and pictures to clarify existing messages. At the outset and throughout the cycle of WNv, the ADH worked to inform, educate, and not alarm the public about WNv. Their early messages involved "analyzing dead birds and . . . getting rid of standing water" (Dr. William Mason, MD, personal communication,

April 7, 2004). In the second year of WNv and with budgeted funds, the ADH distributed 23,000 posters and brochures to the public regarding using Deet and the importance of draining standing water. ADH also printed coloring books for county fairs and schools (Boozman, 2002).

A wide variety of approaches were used in their communication campaign to help the public better understand the risks associated with WNv. ADH had a clear message for the public: Drain standing water, analyze any dead birds, and stay in during dawn and dusk. In doing so, ADH used traditional outlets for their message including posters and brochures. However, it also used coloring books to engage both children and adults in their message. ADH clearly met Covello's (2003) standards for a clear message by using a wide variety of strategies and making the message understandable and engaging. What follows is a summary of effective message strategies for communicating clearly during WNv:

- ADH responded clearly about use of Deet, draining standing water, remaining inside during dawn and dusk hours, and analyzing birds.
- ADH used posters, brochures, and coloring books to clarify their messages.

Meet the Needs of the Media. The media is the primary means for reaching the public. During a crisis, the media provides the immediate access necessary for a rapid response (Reynolds, 2002). Covello (2003) explains that meeting the needs of the media involves being accessible, telling the truth, and providing background about complex risk issues. Dr. Fay Boozman, MD, director of ADH, explained that early on in the crisis the Department of Health was "available to the media every day" (personal communication, April 22, 2004). In addition, Dr. Mason, MD, medical leader for ADH Bioterrorism Preparedness, explained that due to the public interest in WNVs, he did a number of radio, TV, and public presentations, including the following:

- August 5, 2002: Channel 11 (TV)
- September 17, 24, 24: Lecture to physicians
- May 13: Channel 4 (TV)
- May 19: Newspaper interview and Channel 7 (TV)
- August 6, 8, 12, 2003: Radio and TV interviews

Additionally, the two assigned PIOs, Ann Wright and Robert Alvey, conducted telephone interviews daily with media throughout the state. In each case, ADH worked to be available to the media and to use the media as another channel for informing the public about the potential risk of WNv. This is clearly a best practice in risk and health communication and illustrates ADH's commitment to maintaining contact with the public and ensuring it had a consistent message about the risks of WNv. In short, ADH was available to the media and used this stakeholder as a conduit to provide information to other publics.

- The ADH was available to media.
- The ADH used media as another channel to get its message out to the public.

Be Truthful, Honest, Frank, and Open. Communication scholars often advocate for openness, honesty, and truthfulness both as a strategic position and as a value stance (Reynolds, 2002; Sandman, 2000; Seeger, Sellnow, & Ulmer, 2003; Seeger & Ulmer, 2001). Covello (2003) suggests that crisis and risk communicators should disclose risk as soon as possible and not exaggerate or overreassure about potential risk. Lee Clark, ADH performance measures leader and director of the WNv EOC team, explained that "the goal of the messages during WNv was to 'educate and inform—don't alarm'" (personal communication, 4/7/04). Ann Wright, ADH team leader for external communication, explained, "We focused on recommendations. We call these SOCOs 'Single overriding communication objectives.' This is what we are trained to do. Also, we focused on the 3 D's: Deet insect repellent, stay indoors at Dusk/Dawn, and Drain standing water" (personal communication, April 1, 2004).

In addition, ADH worked with at-risk groups' to make sure that they had appropriate information about their level of risk. Ann Wright, team leader for external communication, said "Environmental specialists would talk to nursing homes and retirement homes about the risk of WNv. We would often direct people to our website and CDC's website" (personal communication, April 1, 2004).

As a result, ADH was able to meet their stakeholders objectives by being open and honest with them. The goal of crisis and emergency risk communication is to provide the information that the target audience wants or needs to know to make an informed decision (Reynolds, 2002). Key steps in this process included having a mindset of educating and informing the public not over- or underestimating the risk, using their Single Overriding Communication Objectives (SOCO) and meeting the needs of at-risk groups. Hence, effective risk and crisis communication in this case involved:

- Educating and informing stakeholders
- Encouraging stakeholders to use Deet, stay indoors at dusk and dawn, and drain standing water; and
- Communicating information to at-risk groups.

Listen to People. Part of effective crisis communication involves seeking feedback (Reynolds, 2002; Sandman, 2000). Covello (2003) explains that risk and crisis communicators need to let parties associated with risk be heard and understood. To accomplish this, channels of communication must be established just for listening and answering questions. Dr. Bill Mason, MD, ADH Bioterrorism Preparedness medical leader, explained that a hotline was established in 2002 and was very effective. One of the reason feedback was important was because WNv was thought to be a new exotic disease. As a result, the disease was newsworthy and highly reported by the news media.

Bernie Bevill, ADH head of Radiation Control for Emergency Management, explained that "the Hotline had 4,107 calls in 148 days. The Web page had 12,510 hits in 148 days" (personal communication, April 1, 2004). Clearly, the public was interested in gaining information about WNv and the ADH was successful in providing that information to the public in an accessible manner. However, ADH also made sure it was listening to the public and gaining information that could be provided to a wide audience. In this case, consistent questions received from the hotline were later added to the ADH Web site.

Robert Alvey, ADH crisis communications specialist, said "We were posting updates almost daily on our website" (personal communication, April 1, 2004). In addition, he explained, "We developed a state map and chart that showed activity by county in addition to our press releases. Some days we had more hits on the website than calls to our hotline" (personal communication, April 1, 2004). ADH wanted to make sure it used more than a hotline to gain information from the public. ADH consistently updated its Web site to ensure that the public would have information and would know where to call if they needed additional information. Listening is critical to any risk and crisis communication effort. What follows are two effective strategies that ADH used to manage the WNv event:

- Hotline established to listen to the public and provide risk information.
- A paradigm shift in the method that stakeholders were using to seek information about the WNv—The Web page became more popular over time than the telephone hotline.

Coordinate, Collaborate, and Partner With Other Credible Source

A number of studies have illustrated the importance of establishing partnerships (Reynolds, 2002; Sellnow, Ulmer, & Snider, 1998; Ulmer, 2001). Covello (2003) suggests that coordinating with other organizations is a key to risk and crisis communication success. In handling WNv, ADH worked with many organizations to ensure the safety of their citizens. For instance,

Ann Wright explained, "We are a part of the 'National Public Health Information Coalition (NPHIC).' They have a listserv where members who represent every state and major U.S. city share everything with each other, like fact sheets. Also we referenced the CDC website a lot and CDC would often have conference calls" (personal communication, April 4, 2004). In addition, the Centers of Disease Control (CDC) assisted in identifying appropriate CDC contacts as questions and issues arose. "The CDC assisted Arkansas by sending a team of Epidemiological Intelligence Service professionals to . . . [develop] a database for human, bird, and equine cases" (Boozman, 2002). Clearly, collaboration with a national health organization like CDC was a beneficial resource throughout the WNv outbreak in Arkansas.

In addition, the EOC worked with physicians "through weekly broadcast-facsimile reports to the Arkansas Hospital Association and the Arkansas Medical Society, to submit serum and cerebrospinal fluid specimens from suspected cases of viral meningoencephalitis to ADH" (Fleischauer et al., 2003, p. 94). Coordinating with physicians across the state enabled ADH to keep a close watch on the number of cases of WNv and related diseases in the state. This approach allowed for a seamless flow of communication between ADH and those physicians treating patients with WNv. In short, ADH partners with national and local people and agencies to help support its efforts. These collaborations provided credibility and effectiveness of the overall response. Other state health organizations dealing with risk or crisis would do well to utilize the following best practices to collaborate with credible sources.

- ADH collaborated with the CDC to share information regarding WNv.
- ADH collaborated with physicians in the state to monitor level of WNv outbreak in the state.

Accept People and Organizations as Legitimate Partners

Accepting stakeholders as legitimate partners in risk communication is a key part of crisis preparedness and resolution (Covello, 2003; Reynolds, 2002; Sellnow, Ulmer, & Snider, 1998; Ulmer, 2001). Organizations that involve other stakeholders in decision making about risk and preparation for crisis are likely to benefit from support of those groups during a crisis. As a result of WNv and other disease outbreaks in the state, ADH has continued to offer CERC training classes for other ADH personnel and has expanded its offering to include traditional and nontraditional external partners in Arkansas. In addition, ADH has partnered with the University of Arkansas at Little Rock to add interns to their bioterrorism team to collect data on risk and crisis communication in a wide variety of public health communication situations and provide learning opportunities for the organization. These learning opportunities will hopefully provide the organization and students with lessons regarding further improvement and development in their risk and crisis communication initiatives. A best practice of any organization that must deal with risk and crisis communication is to leverage its resources so it is able to respond to potentially devastating events. ADH has continued to develop relationships within and outside of the state to prepare for and respond to risk and crisis communication. Two clear initiatives in their development of partnerships are as follows:

- Continue with CERC training in the state
- Partner with UALR and the Department of Speech Communication

The ADH's 2002–2003 WNv risk and crisis communication awareness effort was, in large part, a success. This chapter argues that a key part of this success was the ADH's application of best practices of risk and crisis communication. Much of the credit should go to those health officials who completed training in risk and crisis communication prior to the WNv event. In our interviews many respondents suggested it was their education in risk and crisis communication including CERC training done by the CDC that led to a better understanding of how to communicate risk to the public. A critique of ADH's response to the crisis suggests that it responded appropriately and effectively in terms of the best practices outlined in this chapter. It appears that ADH will continue to improve its communication and education in risk communication as it forms partnerships and continues to evaluate its communication efforts in other risk and crisis scenarios. This chapter concludes with implications regarding how ADH plans on building on its risk and crisis communication capabilities in the future.

IMPLICATIONS FOR PUBLIC HEALTH RISK AND CRISIS COMMUNICATION: WHAT IS IN THE FUTURE FOR ARKANSAS DEPARTMENT OF HEALTH

This case analysis of WNv in Arkansas examined the effectiveness of communication in addressing a novel, exotic virus outbreak. Lessons learned in this case suggest that effective risk and crisis communication principles are useful benchmarks for health departments striving to improve their public health communication. In particular, Covello's seven best practices are useful in assessing the effectiveness of communication challenges faced by public health. Moreover, the CDC is achieving success in its efforts to prepare health departments effectively to communicate risk to their publics as well as managing the potential crises that any state may face.

What follows are some key issues that the ADH is using to learn from WNv and to build on their successful WNv public risk and crisis communication efforts.

- Continue to scan the environment for potential crises.
- Establish coalitions with neighboring states on risk and crisis preparation.
- Pursue annual conference on risk and crisis communication that connects practitioners and scholars.
- Develop risk and crisis communication training for media.

WNv was a wake-up call for public health in many of the same ways as was the anthrax episode. WNv illustrated that an unanticipated new emerging infectious disease could spread rapidly, and that public health might have limited ability to control an outbreak. Although disease monitoring has been a quintessential feature of public health for some time, WNv reiterated the need for both rapid recognition and rapid public warning. Moreover, this outbreak illustrated again that disease does not respect state or even national borders. Public health must coordinate and cooperate across state lines and at the federal level. Learning can be facilitated by activities such as national conferences, particularly those that bring together the research and practitioners communities and published reports regarding response effectiveness. Finally, ADH learned again that the media are an important public health partner during a crisis. Without the assistance and cooperation of the media, the public would be uninformed or poorly informed. Facilitating the effectiveness of this communication requires that public health and media understand one another and that both appreciate the unique exigencies of risk and crisis communication.

Initiatives such as these provide ADH and other state and local departments of public health with continual improvement of their risk and crisis communication competencies. In addition, these new initiatives illustrate the willingness of ADH to learn from both successes and failures and continually improve their risk and crisis communication. Clearly, the best practices described here are useful goals for any organization or agency facing the risk of a crisis. They are not, however, a panacea. Rather, they are part of a process of ongoing improvement. A number of important issues remain, including building coalitions with neighboring states and federal agencies, facilitating opportunities for vicarious learning from other crises and disasters, and developing stronger partnerships with the media.

REFERENCES

Boozman, F. W. (2002). *West Nile Virus testimony*. U.S. Senate.

Coombs, W. T. (1999a). Information and compassion in crisis responses: A test of their effects. *Journal of Public Relations Research, 11*, 125–142.

Coombs, W. T. (1999b). *Ongoing crisis communication: Planning, managing, and responding*. Thousand Oaks, CA: Sage.

Covello, V. (2003). Best practices in public health risk and crisis communication. *Journal of Health Communication, 8*, 5–8.

Covello, V., Peters, R., Wojtecki, J., & Hyde, R. (2001). Risk communication, the West Nile Virus epidemic, and bioterrorism: Responding to the communication challenges posed by the intentional and unintentional release of a pathogen in an urban setting. *Journal of Urban Health. Bulletin of the New York Academy of Medicine, 78*, 382–391.

Covello, V. T. (1992). Risk communication: An emerging area of health communication research. In S. A. Deetz (Ed.), *Communication Yearbook 15* (pp. 359–373). Newbury Park, CA: Sage.

Fink, S. (1986). *Crisis management: Planning for the inevitable*. New York: AMACOM.

Fleischauer, A. A., Williams, S., O'Leary, D., McChesney, T., Mason, W., Falk, S. et al. (2003). The west nile virus epidemic in Arkansas, 2002: The Arkansas department of health response. *The Journal, 100*, 94–99.

Pauchant, T. C., & Mitroff, I. I. (1992). *Transforming the crisis-prone organization*. San Francisco: Jossey-Bass.

Penrose, J. M. (2000). The role of perception in crisis planning. *Public Relations Review, 26*, 155–171.

Reynolds, B. (2002). *Crisis and emergency risk communication*. Atlanta, GA: Centers for Disease Control and Prevention.

Reynolds, B., & Seeger, M. W. (2005). Crisis and emergency risk communication as an integrative framework. *Journal of Health Communication, 10*.

Sandman, P. (2000). Open communication. In E. Mather, P. Stewart, & T. Ten Eyck (Eds.), *Risk communication in food safety: Motivating and building trust*. East Lansing, MI: National Food Safety and Toxicology Center, Michigan State University.

Seeger, M. W., Sellnow, T. L., & Ulmer, R. R. (2003). *Communication and organizational crisis*. Westport, CT: Praeger.

Seeger, M. W., & Ulmer, R. R. (2001). Virtuous responses to organizational crisis: Aaron Feuerstein and Milt Cole. *Journal of Business Ethics, 31*, 369–376.

Sellnow, T. L., Seeger, M. W., & Ulmer, R. R. (2005). Constructing the "new normal" through post crisis discourse. In D. O' Hair (Ed.), *Community preparedness and response to terrorism: Communication and the media*. Westport, CT: Praeger.

Sellnow, T. L., Ulmer, R. R., & Snider, M. (1998). The compatibility of corrective action in organizational crisis communication. *Communication Quarterly, 46*, 60–74.

Ulmer, R. R. (2001). Effective crisis management through established stakeholder relationships: Malden Mills as a case study. *Management Communication Quarterly, 14*, 590–615.

Vanderford, M. (2003). Communication lessons learned in the emergency operations center during the CDC's anthrax response: A commentary. *Journal of Health Communication, 8*, 11–12.

Yee, D. (2002, August 9). $1 million available to thwart West Nile Governor offers counties emergency aid. *Arkansas Democrat-Gazette,* p. A1.

7

LESS AMBIGUITY, MORE HOPE

The Use of Narrative in Chinese Newspaper Reports on the SARS Crisis

Min Liu

J. J. McIntyre

Timothy Sellnow

The initial messages shared by the Chinese Government grossly understated the SARS epidemic. As the disease spread, the government faced the rhetorical challenge of reestablishing its legitimacy with the Chinese public. To do so, the Chinese newspapers included an evolving narrative of health care "heroes" who had helped contain the disease and develop appropriate care. This chapter analyzes the heroic narratives from the perspective of communicating control in the crisis situation.

Crises are typically characterized by an escalating threat, short response time, and surprise (Seeger, Sellnow, & Ulmer, 2003), which are further complicated by the involvement of various parties possessing different and often opposing perspectives, interests, and motivations. Varying interpretations of crisis situations by stakeholders create multiple understandings of an event or a series of events and, as a result, manifest in the creation of competing versions of the same crisis. The contrasting stories evolving from a crisis can be understood as narratives (Hay, 1995). The narratives or stories are forced to compete with each other for stakeholder acceptance. Stakeholders searching for order and understanding in a crisis must judge the competing narratives and derive their opinions based on their overarching knowledge about the situation. The narrative perspective explains the communication strategies within a crisis situation, which are often used to help construct a favorable social reality for the audiences (Venette, Sellnow, & Lang, 2003). This chapter examines the use of such narratives created by the Chinese government to restore legitimacy damaged by the government's initial handling of the SARS epidemic.

CRISIS COMMUNICATION

Heath and Millar (2004) suggest that a crisis has two dimensions. The first dimension is the technical/managerial, which recognizes that a crisis is created by the managerial choices and technical performances of an organization. The second dimension is the communication response before, during, and after the crisis. This study focuses on the latter dimension. First, this section discusses the importance of effective communication during a crisis and examines various stakeholder responses to crisis communication. It then explores the organizational benefits to effective crisis communication: how crisis communication can be framed within a narrative perspective and how an organization can use narrative to create an effective crisis communication strategy.

Effective communication is essential to an organization's recovery from a crisis. Not only can crisis communication help control and bring an end to a crisis, it can even create a more positive reputation for the organization involved in the crisis (Seeger, Sellnow, & Ulmer, 2003). NASA was successful in maintaining public trust and confidence in its vigilant and strategic handling of the Apollo 13 crisis (Kauffman, 2001). Nike was also successful in its image-restoration strategy. The company reestablished its social legitimacy by creating a public narrative that emphasized its ongoing effort to improve working conditions in developing countries worldwide (Sellnow & Brand, 2001). Additionally, Schwan's maintained its social legitimacy through the largest common vehicle salmonella outbreak in history by creating a crisis narrative that depicted the company's social responsibility for the outbreak. Using corrective action, Schwan's turned a potentially fatal crisis into a demonstration of the company's values and customer commitment (Sellnow, Ulmer, & Snider, 1998). These success stories serve as examples of the importance of framing a positive narrative to stakeholders.

The threatening nature of a crisis can be extremely overwhelming to stakeholders. A sense of hopelessness and helplessness may develop and become more debilitating than the actual threat. Thus, it is crucial that early messages from a designated spokesperson respond to the crisis in a way "that empowers others" (Barrett et al., 2003). The initial communication should provide information about potential threats and mitigation activities. The discussions create an initial frame of the crisis and enact sensemaking in ways that facilitate a return to normalcy and control or at least a feeling of such. Sensemaking is a way to create meaning by collaborative agreement in a context intended to reduce confusion (Weick, 1993). "Thus, organizations become important because they can provide meaning and order in the face of environments that impose ill-defined, contradictory demands" (p. 635).

Heath and Millar (2004) contend that a crisis can be viewed as a struggle for control. People who are affected by a crisis look to the responsible party for actions that will control the situation and reduce harm. Heath and Millar point out that the central focus in crisis communication should be a theme of responsibility. The organization must rise to the rhetorical exigency by taking action and effectively communicating to the public about its efforts to regain control of the circumstances. Messages sent by the organization need to "responsibly define the crisis, the actions that need to be taken, the actions that will be taken, are being taken, or have been taken" (p. 9). The messages may also contain information that stakeholders use to form their understanding of the developing crisis narrative.

As Seeger, Sellnow, and Ulmer (2003) write, crises, if managed effectively, can be an opportunity to create fundamentally new knowledge and understanding among its members. A severe crisis "may prompt a systematic search for a new knowledge base" (p. 81). The initiated search prompts a process of selection and enactment that allows organizational members to "collectively construct this new knowledge base" (p. 81).

Some crisis communication researchers have adopted a rhetorical perspective to understand how organizations create narratives and tell their stories in response to crisis circumstances (Benoit, 1995a, 1995b; Hearit, 1995a, 1995b). Heath (2004) posits that a crisis is an interruption of a narrative—the normal activities of an organization—and it also begins its own narrative, one of uncertainty and ambiguity, one that may or may not end with a return to a normal set of circumstances like those held prior to the crisis.

Narratives change over time, and individuals are often confronted with competing accounts of the same event. Heath (2004) explains:

> Narratives are a way of thinking, a way of ordering the events of the world which would otherwise seem unpredictable or incoherent. As people order the events of their world into meaningful patterns, those patterns-scripts and themes-express their values and guide their actions. However, not everyone thinks or acts in the same way. Viewed in this

> way, narrative analysis assumes that people choose among competing stories that account for a given event. Thus, the question is often not whether the narrative, but which account is best. (p. 171)

Organizations working proactively to create the publicly accepted crisis narrative must "present a factually accurate, coherent, and probable account for the event and its proper resolution" (Heath, 2004, p. 168). The narrative of a crisis will likely undergo changes throughout the crisis because the narrative's meaning is ultimately "co-created by the organization, many commentators, and key publics who seek a rational explanation of the event" (p. 168).

In this view, a crisis creates a rhetorical situation for which the organization needs to enact a narrative as a rhetorical response. In crafting the response, an organization's narrative is comprised of the following: a scene, personae, and plots and themes. The elements are part of the organization's ongoing story; thus, "each of these factors becomes part of the rhetorical problem and therefore requires responses by the organization" (Heath, 2004, p. 11). An organization's response to a crisis is "a narrative framed by scene, moved by personae, in which a theme is created and a plot is enacted" (p. 12).

The element of scene asks and answers these questions: *Where did the crisis occur?* and *What is the symbolic power of the scene?* The personae element of a crisis narrative refers to the persons and factors that led to, occurred during, and transpired to resolution after a crisis. Theme focuses on the central idea of the story and asks the question, "Will the organization's thematic account of the crisis—cause, duration, and magnitude—sustain itself against the scrutiny of other interested parties?" (Heath & Millar, 2004, p. 12). It is important to note that a single narrative does not have to emerge from the organization facing a crisis. Themes of individual stories can form a synergy through continuity that supports or admonishes the overall plot (the story line that the organization wishes to feature).

The final element of a narrative response, again according to Heath (2004), is personae. Personae may include individuals such as heroes or heroines—persons and factors that attenuated, contained, and restored operations so that the crisis did as little harm as possible. In addition, personae may include allies—persons who assisted the heroes or heroines. Collectively, the elements of narrative response provide organizations the means to construct a crisis response that resonates with stakeholders.

Organizations seek to convey a favorable narrative to its stakeholders that exhibits control over the circumstances. The best-case scenario in a crisis, according to Heath (1997), is for the organization to exert control over activities in ways that ensure stakeholders and stakeseekers (key publics) that their interests are cared for by the organization. Narratives accomplishing this task possess a calming quality because, as Heath (2004) explains, "People desire order and predictability that lead to positive rather than negative outcomes" (p. 167).

In this study of the SARS epidemic, we are particularly interested in the elements of the narrative response. Specifically, we focus on how the Chinese government developed a narrative of heroism that sought to restore calm and reestablish public confidence through an explanation of how the doctors and nurses in the country were gaining control over the outbreak. Next, the study provides a summary of the SARS outbreak situation, which is followed by an analysis of the print coverage of the SARS epidemic in the *People's Daily* (the major Chinese newspaper). Finally, we offer a series of conclusions and implications.

THE SARS OUTBREAK

In early 2003, during the outbreak of SARS in Southern China, the Chinese government, faced with pressure from both its domestic and international stakeholders, needed to respond to the public health epidemic with an immediate and effective crisis response. Instead, the government's response during the onset of the epidemic in 2003 was characterized by uncertainty as to what actions to take and what outcomes to expect coupled with an urgent need to avoid public panic.

As a result of its infamous use of contrived statistics to create a false appearance of safety and control in the early stage of the crisis, the government as a crisis communication rhetor had unfortunately suffered from a damaged persona or one of villain (Heath 2004). Thus, the Chinese government was forced to address or intensify the crisis situation after it had damaged its own credibility.

Precrisis and Crisis Communication: Exigency for Control

The technical dimension of the accusation is that the government failed to contain the disease effectively; comparatively, however, the communication failure was much more offensive because it was believed to be directly correlated with the loss of confidence in the Chinese's crisis response (see Table 7.1). The turning point that led to the legitimization of the crisis began on April 8 and 9, when accusations were made by the global community about the Chinese government withholding information (CNN, 2003).

While the outbreak was expanding at an exponential rate, the government was failing on another battlefield—the charges that the government was irresponsible and unethical, causing public confidence and trust to drop dramatically (e.g., Bezlova, 2003). The Chinese public was aware of the criticisms coming from outside the country; however, the Chinese government did not publicly recognize these charges.

On April 8, a military physician in Beijing issued a statement to world media revealing that the numbers of patients infected with SARS were significantly higher than those made public by the Chinese government, which altered the way the world and later China responded to the crisis (*Time*, 2003). At the April 11, 2004, World Health Organization (WHO) press conference on SARS, the WHO executive director of Communicable Diseases said that WHO officials had heard conflicting reports from health workers in China, and that what was occurring in Beijing was more serious than what the government had told the world (World Health Organization, 2003).

Although criticisms from outside the country were seldom reported by the government-controlled news agencies, throughout the crisis, the Chinese people had access to e-mail, instant messaging (IM), Internet, and other communication channels. The access to other forms of mainstream news media kept the Chinese people in touch with the outside world. During the early months of the 2003 outbreak, cell phone users in Beijing

TABLE 6.1.
Timeline of the SARS Outbreak

DATE	TIMELINE OF EVENTS
March 21, 2003	Chinese government asked WHO for help investigating the outbreak of SARS in Guangdong province. A team of WHO experts traveled to the region.
April 7	WHO issued a recommendation that people consider postponing all nonessential travel to Hong Kong and Guangdong province.
April 8	Doctors in China said there were more SARS cases than the government was reporting.
April 9	WHO said China might be withholding information about SARS.
April 18	China responded to the criticism about how it was handling the SARS epidemic by being more open about cases.
April 20	China, under fire for not disclosing the extent of its SARS infections, raised its number of cases from 1,512 to 1,807. The health minister and the mayor of Beijing were dismissed from their posts.

Source: CNN. Timeline: SARS outbreak. April 24, 2003.

alone sent out 7 billion IMs, an increase that made the telecommunication provider industry one of the few that actually gained business in the crisis (PCOnline, 2003). The Chinese people turned to such nonmainstream media as an alternative to obtaining SARS-related information from the government-controlled mainstream media. As a result, the Chinese government was losing control of the public narrative regarding SARS. One means used by the Chinese government to regain control over the crisis narrative was its coverage of the crisis in the *People's Daily* newspaper.

METHOD

The data used for this chapter included news articles from the Chinese newspaper, *People's Daily*. This newspaper was selected because of its unique status among the Chinese media outlets as the party newspaper. *People's Daily* (n.d.) and its subsidiary papers at the provincial level are run by the committee of the Communist Party of China and, as stated in its Web site, is the "most influential and authoritative newspaper in China." The articles were retrieved from the newspaper's Web site at <http://www.snweb.cn/gb /people_daily/>. The Web site was sponsored by the Information Center of *People's Daily*.

For the purpose of this study, only the news articles published in the 37 days from April 5, 2003 to May 14, 2003 were collected for investigation. These days were selected for analysis because as the epidemic developed into an acute stage, SARS-related reports by all major Chinese newspapers peaked from early April to mid-May (Zhang, 2003). News on April 20, 2001, was not made available on the Web site for reasons unknown to the researchers and therefore was not included in the analysis. Altogether 1,122 news articles were included in the analysis.

CODING PROCESS

The following criteria for coding narratives versus non-narratives were established and used by two independent coders to obtain data eligible for the study:

- To be categorized as narratives, the news story must have all the characteristics of a narrative story: a plot; characters; action; a sequence of beginning, middle, and end; and a climax (Fisher, 1987).

- Stories had to focus on individual(s) and organization(s) with regard to their unique experiences in the war against SARS.
- Government officials, including government leaders and various governmental bureaus/offices, are frequently reported with regard to their efforts, stances, and behaviors in coping with the crisis; however, these reports were excluded in the coding. In other words, only the stories from nongovernmental perspectives were used in the final theme analysis.
- To clarify the coding process, coders were asked to pay close attention to news titles that included the following words, the literal meanings of which have a connotation of narrative and/or stories:

    I. *Jishi* (*adversaria*)
    II. *Fang* (visited) *Zaifang* (revisited)

Using the above criteria, the researchers conducted an initial trial round with coders using the news on *People's Daily* from July 17, 2003 to July 20, 2003. During the initial coding, the two coders arrived at consensus on over 85% of the stories coded for narrative versus non-narratives and 90% consensus on governmental versus non-governmental perspectives. Where consensus was not reached by the coders, the coding process was resolved by the researchers.

NARRATIVES OF CONTROL

Early *People's Daily* reports on SARS had a positive and optimistic tone. From April 4 to April 10, there were an average of five daily reports on SARS; by April 11, the reports dramatically increased to 12. The majority of the reports were from government officials and organizations such as hospitals. The news reports focused on transportation and logistical issues, illustrating people working with good efficiency to handle the disease (e.g., "China Was," 2003). Also noticeable was the use of stories about patients who recovered and doctors who were confident about their ability to deal with the disease. For example, on April 9 and 10, the *People's Daily* featured multiple reports of optimism, including stories of patients who had recovered from the disease in Guangdong ("Guangdong Convalesced," 2003; "Guangdong Hospital Visited," 2003) and Hong Kong ("Hong Kong 138 Cases," 2003). Another report on April 10 included pictures of two healthy-looking, smiling patients, who "have recently been infected with SARS, and as the majority of SARS patients in the hospital, have been cured of the disease and will be discharged from the hospital soon" ("SARS Patient Recovered," 2003, para 2).

The storytelling took an interesting turn 2 days later, on April 12, when a detailed interview with the Guangdong Disease Control Center attempted to respond to attacks on the validity of SARS-related statistics, as shown in the title "From Guangdong—How Are the SARS Data Collected and Is It Reliable?" ("From Guangdong, 2003). This article included detailed information about how disease control center personnel worked and how SARS case data had been collected in the Guangdong province. The three subtitles are "Data Collection in Guangdong Started at the End of 2002" (para. 2), "We Established a Database System and Tracked Down Every New Case" (para. 4), and "WHO Expert Team Found Guangdong's Data True, Believable, and Reliable" (para. 10).

Apparently, educational information about how the disease control center functions was not enough to calm concerns. From this turning point on April 12, 2003, the *People's Daily* started a different approach to SARS reporting and enacted a new narrative about the crisis. The rhetorical response to the crisis exigency for control was the enactment of a narrative that relied heavily on the use of heroism.

BUILDING OF THE HEROISM PERSONAE

In a rhetorical crisis response, the rhetor tries to provide a competing narrative by telling a story that hopefully will lead to a resolution of the crisis. Personae refer to the cast of characters in such a narrative. According to Heath (2004), personae in a crisis narrative take the forms of heroes/heroines, enemies, allies, and so on. In *People's Daily's* SARS narrative, a metapersonae of medical workers were portrayed as heroes in the war against SARS. The heroes were created to convey a narrative of control and hope. Specifically, stories were told using themes of sacrifice, professionalism, fearlessness, selfless commitment, and optimistic morale. The themes formed a synergy that supported the overall plot that China is in good hands with the most committed and professional medical workers, and that there should be more hope and less panic because the people have the outbreak under control. The following discussion is based on the emergent character elements that were used to define and depict the heroic personae—the evolving themes of the SARS narrative.

SACRIFICE

Medical professionals were depicted as willing and never hesitant to sacrifice for the patients. In fact, hospital workers were characterized as sacrificing their families for their work to fight SARS. Several stories were told about young medical workers leaving newlywed spouses, ill children, or aged parents to work in the demanding environment of quarantine. An article on May 2, 2003, told the story of a Dr. Qin and his family who sacrificed with him through the crisis ("United We," 2003):

> Dr. Qin also has important responsibilities as a doctor-in-charge on SARS. Ever since the end of March he hasn't been back at home for more than 4 weeks. Taking care of his daughter who goes to school and his sick dad all falls on the shoulder of Qin's aged mum. When he left, his mum said, "go to your work and don't worry about us. I will take care of us back at home." It was not until recently his parents got to see their son for the first time, but it was on TV, when he was all covered up with the exposure unit. (para. 10)

Dr. Zhang, from Tongren Hospital, was in charge of calling together the team to work on the frontlines against SARS. What surprised Zhang is how supportive and understanding the doctors' and nurses' parents and families were when they heard from him that their sons and daughters—in most cases, the major caretakers in the families—were to work on the most dangerous SARS frontline. Zhang said, "When I talked with them over the phone and heard them being so understanding, using the most humble and simple language, I just couldn't help but cry" ("United We," 2003, para. 8).

Another sacrifice plot is that, in this war, medical workers developed camaraderie and sacrificed for each other. Workers took care of each other by constantly reminding their colleagues to take preventive medication, to eat, and to rest well. The medical teams supported and thought for each other just like warriors fighting shoulder to shoulder. They knew that each of them needed to stay healthy and safe for each other and for the victory of this brutal war. The following story depicts an emotional moment that nurses shared when two of their colleagues fell prey to the disease:

> Nurse Liu said with tears in her eyes that "what you journalists should interview most are not us who are still healthy, but those sisters who've been brought down by SARS now. Nurse Deng and Nurse-in-Charge Liang were unfortunately infected; I went to visit them with other sisters but they kept pushing us away and out of the ward, because they didn't want to risk infecting us too. After they were sent to the wards, they were crying inside the glass window and we were crying outside. They asked us to leave them with the medicine and everything; they wanted to inject and take the medicine all by themselves so to protect us as much as possible." ("Gods of Protection," 2003, para. 12)

The following story describes a patient's heartfelt appreciation of a doctor's selfless sacrifice: A patient about to recover and leave the hospital wrote in a letter to Dr. Jiang Rongmeng,

> I remember being in high fever I was lying there, heavy-handed and dizzy, but then I heard someone coming in and greeting me, asking me how I was feeling. I couldn't see who he was with the layers of protection clothes, but I could recognize the voice, it's Dr. Jiang. I also knew he had not left the quarantine wards for at least 48 hours. Looking at him my heart hurt so much; at that moment I was wishing all of us could get better j;ust so that our doctors and nurses could at least go home for their families and get some rest. ("Building the Line," 2003, para. 15)

PROFESSIONALISM

Stories were also told about the professionalism of these medical heroes. They never doubted their responsibilities as medical workers; they were able to make the right decision although it was their own life that was at stake. They performed their routine operations with great rigor and excellence, despite that they might be the next person coming down with the outbreak virus. On May 7, a moment of professionalism and fearless sacrifice was captured in the following story of Nurse Zhong:

> A day a suspect SARS patient who had had continuous fever was sent to the hospital; he was too weak to walk by himself; Nurse Zhong with others carried him into the ward though everyone knew the risk it involved. The patient was very touched and said, "Even my family wouldn't want to stay this close to me; aren't you afraid of the disease?" Zhong, who was very humble with few words, said, "Yes we are afraid too, but this is our responsibility." ("Gods of Protection," 2003, para. 16)

Even when they had been infected, the doctors and nurses still did not forget who they really were and did everything they could to help their colleagues to diagnose; they kept record of their own symptoms in the most professional way and asked for the newest medication so as to provide precious data for testing. On May 4, 2003, the story of the first military medical worker killed in this battle, Dr. Li Xiaohong from Beijing Police Hospital, was reported in great length:

> After the diagnosing, her managers, colleagues and family all worried very much about her, but Li herself seemed exceptionally calm and

> peaceful. She didn't want her case to lead to any unnecessary panic among the colleagues in the hospital and among the patients, which would break down the normal routine practice in the hospital. She told others, "I am a doctor; trust me I know how to take care of myself."
>
> April 6: things got worse for Li; she started to have problems speaking; every sentence she tried came with a painful shortness of breath. The management in the hospital decided to transfer her to the PLA 301 Hospital where there were better facilities. After she knew this she managed to say, "I am still young and have stronger immunity; let's transfer the other patients who are in more difficult conditions!"
>
> Later Li was transferred to PLA 302 Hospital; however, her condition didn't show any better signs. The management decided to let her husband, also a doctor, stay with her and take care of her, but she didn't agree. She said to her husband, "The hospital is short of people; I am already lying here doing nothing; I can't drag you here too. You should go back to your work!" ("Fighting Against SARS," 2003, paras. 8–10)

When her moment came, her final words at the deathbed were still notes of self-diagnosis and words of encouragement for the colleagues she left behind:

> The day came when doctors announced her [Dr. Li] in serious condition. She could no longer utter her thoughts, but with a surprisingly strong will she still managed to communicate with doctors in writing her reactions and analyses of herself, even it's though she could barely get through a few words. ("Fighting Against SARS," 2003, para. 16)

FEARLESS AND SELFLESS COMMITMENT TO THE PATIENTS

During late April and early May, the disease spread rampantly. Within a week, the confirmed cases increased from 2,427 on April 27 to 3,799 on May 2. During the acute stage of technical dimension of the crisis, messages sent out by the government seemed to take a different focus as well. Specifically, a new theme was added to the storytelling done by the government. Stories were now told about how medical heroes were extremely committed to the safety of their patients, and the stories described the fearless and selfless actions of these heroes when they were faced with the most feared and least known outbreak to date. The first few cases of causalities in hospitals were reported extensively. A news article, on April 26 featured the first report of medical workers killed in this battle ("The epical story," 2003). On May 4, the newspaper feature detailed reports of the stories of two doctors who died on their posts in the frontline (An epic of life," 2003; "Fighting against SARS," 2003; ").

These and other stories depict fearless wartime hero personae who never think for the safety of themselves and who are always ready to sacrifice themselves for the safety of patients and colleagues. The following is an excerpt from the May 4 story of Dr. Liang ("The epical story," 2003).

> Xie, the new patient was just confirmed to have SARS, the vicious epidemic that this hospital had not dealt with until then. Also it was known that the patient's only son had just died after being infected by her. Liang without any hesitance signed his name as the doctor-in-charge for her. After Liang's death, nurse Ma Jie recalled, "Every time when we had such dangerous cases, Dr. Liang would always sign himself as the doctor-in-charge; it's almost as if we had been used to it. Now I wish I had stopped that." ("An epic of life," 2003, para. 4)

In the report, Dr. Liang's classmate in medical school remembers calling him not long before that, warning him to be cautious with SARS patients, but Liang's response was that being a doctor means that he should not and would not put his own health before the patients. The story continues,

> April 26th, patient Xie was in most critical condition, and the only way to save her life was to cut through her respiratory tract and apply oxygen apparatus. Anyone with the basic medical knowledge of respiratory disease knows that once a patient's respiratory tract is cut open, immediately all the infectious elements will outpour in great volume, and everyone present will very likely be affected. However the emergent circumstance could not afford to wait. Dr. Liang started early that day to prepare for the surgery. When the surgery finally took place as planned, Liang and the other 7 doctors and nurses present were doing every step of the process with perfect professional caution and excellence. Not a moment did they hesitate or worry about how dangerous it was. What nobody expected is that it is just this operation that got Liang infected. ("An Epic of Life," 2003, paras. 5–7)

The previous example illustrates how the first groups of medical workers acted while working on the frontline. They were perfectly aware of the enormous danger involved; however, they had literally no alternative resort to control and treat the disease safely. Their professional training taught them what precautions to take, but many of them did not let the fear of uncertainty slow down their pace. The story of Dr. Deng Lianxian, who was one of the few doctors who died from treating the early cases was also reported in great detail ("The Epical Story," 2003).

OPTIMISTIC MORALE

Another theme parallel to the previous themes runs through the whole storyline of heroism; that is, the theme of optimistic spirit and confidence. In addition to the details of the overwhelming daily work and the emotional moments of loss and memories, there were also detailed reports of how medical workers remained upbeat and optimistic during the most nerve-breaking moments in their daily work routine, like the following story that appeared on May 7:

> "Happy birthday to you . . . " the night of May 5th, one could hear the melody of the birthday song from somewhere in Xiaotangshan SARS Hospital. People were having a birthday party during a brief break from the battle against SARS. It's the 30th birthday of nurse Yang Yanping, who joined the SARS hospital from No. 304 Military Hospital.
>
> Though the past few days had brought in three batches of patients, all the overwhelming work and immense pressure did not take away the happy and positive spirit from these nurses. Early in the morning they ordered a birthday cake for Yang, and also prepared for a few fun tricks. Ten other nurses who came from another hospital also prepared a great gift for her. Though they were from different hospitals and never knew each other until they joined Xiaotangshan SARS Hospital, this battle against SARS brought them together as a team, and they felt more like family to each other rather than just colleagues. ("Birthday Song at," 2003, paras. 1–2)

The story ended with this note: "Two hours later, the party was over and the girls started another night's hard work while still humming the songs. Tonight they were expecting 90 patients" ("Birthday Song," 2003, para. 5).

CONCLUSION

During the acute stage of the SARS crisis, the Chinese government was losing control of the crisis narrative to the external information sources in the international community (e.g.,*Time*, 2003; WHO; 2003). Competing narratives damaged the public's trust in the Chinese Government's ability to bring the epidemic under control. In response to this communication exigency, the Chinese government (as the crisis communication rhetor) enacted a narrative of heroism, consisting of heroic stories of the health care workers who helped contain the disease and develop appropriate care. Through *People's*

Daily, these messages reached the general audiences, helped reduce uncertainty and fear, and communicated control and hope in the acute crisis stage.

The themes presented in the overall narrative told an epic story of medical heroes who worked fearlessly on the frontline of the war against SARS. The heroes never hesitated, even when their own lives were at risk, and consistently chose patients' safety over their own. The heroes worked day in and day out both treating patients and trying to find a cure. The heroic narrative storyline was constructed with vivid details using the eyes and voices of actual witnesses such as patients, journalists, peer doctors, and nurses.

The strategic choice of the frontline doctors and nurses as the personae for the heroism narrative is crucial for the credibility of the message. The unique status of medical professionals in the context of a national health crisis gave credibility to the stories; specifically, the accounts of selfless devotion helped build a convincing character while depicting their medical expertise. The combination of success and professionalism of medical professionals displayed a strong message of China's competency in confronting the epidemic.

As the stories unfolded throughout the crisis, the focus shifted more to the life-changing moments of the medical workers involved; the stories display more personal sacrifice and considerable success in the war against SARS. Other themes that emerged included what the selfless work meant to medical professionals, how their family's understanding went above and beyond normal circumstances, and the efficiency medical workers demonstrated in constructing a quarantine area for patients.

Another important factor in the development of the heroism narrative is the highly repetitive coverage of the medical hero stories. The sheer amount and depth of coverage of the medical heroes is unprecedented for the *People's Daily*. The *People's Daily's* is the only comprehensive national newspaper in China, and its normal reporting style emphasizes breadth over depth. The in-depth, emotional, and highly repetitive reporting of medical hero stories gives an increased importance to the storyline and emphasizes a message of hope to the Chinese people.

The language used in the hero stories played a significant role in recognizing the importance of medical professionals. The stories used a special register composed of wartime vocabularies that are intense and emotionally charged. Examples include phrases such as "warriors who fight on the frontline" ("Gods of Protection," 2003, para. 18) and "to win the battle against SARS" ("Birthday Song at," 2003, para. 3). Medical personnel were characterized as "wartime heroes" fighting "to build a line of defense" in this "real war that does not use bullets" ("Gods of Protection," 2003, para. 18). The medical workers who initially came into contact with the disease were described as "the warriors who charged on the frontline of the SARS" ("Fighting Against SARS," 2003, para. 1) and "land miners" who cleared away the uncertainties and provided the most precious knowledge and "ensured the final victory in this battle" ("An Epic of Life," 2003, para. 17).

Language also placed the narrative's heroes in a glorious historical context. More than once the title *the most loveable people* was used to refer to the doctors and nurses fighting against SARS ("Gods of Protection," 2003, para. 18). This label has a special meaning in the Chinese culture. The title belonged to the people's collective wartime memory of the 1950s and, up to this point, had always been reserved for military heroes. The SARS news discourse used the special connotation of this phrase and redefined the crisis experience for the general audience.

IMPLICATIONS

This study offers a glimpse into the use and construction of crisis narratives. This section discusses how the Chinese government's initial response to the crisis endangered its legitimacy and the difficulties of creating a narrative response to a crisis, and it offers the reader some successful attributes of the SARS crisis narrative. Finally, this study posits an extension to Heath's theory of crisis response.

According to Heath (2004), "A crisis event constitutes a rhetorical exigency that requires one or more responsible parties to enact control in the face of uncertainty in an effort to win key publics' confidence and meet their ethical standards" (p. 167). The Chinese government's initial response to the SARS epidemic downplayed the seriousness of the threat by misrepresenting the number of cases and the seriousness of the epidemic. Heath and Millar (2004) argue that over- or downplaying the crisis could lead to serious consequences. In practice, the Chinese government's initial misrepresentation of the facts surrounding the crisis did cause a struggle that threatened the legitimacy of the government on local and global levels. Various internal and external stakeholder groups faced with incongruent statements about SARS began to construct alternative narratives in a "search for order" (Heath, 2004, p. 167). The Chinese government then had to overcome these competing narratives to regain legitimacy by displaying character and competence in handling the crisis exigency.

An organization's response to a crisis does not exist in a vacuum. "The narrative of each crisis results in meaning co-created by the organization, many commentators, and key publics who seek a rational explanation of the event" (Heath, 2004, p. 168). Factors that an organization needs to consider when constructing a response to an exigency include the needs of the key stakeholders, the competing narratives, the initial narratives of the organization, and the stakeholders' reactions to the initial stories. The crisis response must display continuity with what is already known by stakeholders to regain the perception that the organization is in control of the situation. "People desire order and predictability that lead to positive rather than negative outcomes" (Heath, 2004, p. 167).

To build a convincing narrative that works to regain control and restore legitimacy, organizations must choose the right personae to convey competence and credibility. The Chinese government, via the *People's Daily's* coverage, selected medical professionals to assume a heroism personae role. In this situation, medical professionals possess and communicate both character and competency; however, the true strength using the professionals is that they have a sustained credibility within the organization facing a battle of legitimacy.

Including local knowledge of multiple stakeholder groups in the response narratives increased the continuity of the message and may have aided the narrative's credibility to different stakeholder groups. The criticism against the government and the newspaper was that SARS-related communication had been untruthful. Thus, the narratives must be perceived as believable to the doubtful public. The *People's Daily* included local information that a variety of stakeholders could relate to and sees as true. The newspaper articles used interviews with real people (citing actual names and direct quotations), coupled with a strategic use of diverse hospital locations that were well known to the audiences in infected areas such as Beijing and Guangzhou.

Also noteworthy was the use of strategic ambiguity that was displayed by the government newspaper's avoidance of the outside world's criticisms. Although the external criticisms did spur a change in both how the Chinese government dealt with the outbreak and the *People's Daily* reporting style, the competing narratives were not recognized or given a voice. The newspaper focused on providing narratives of hope to help regain legitimacy and rebuild confidence; however, the newspaper did not directly acknowledge or attempt to discredit the competing stories. Instead, stakeholders were allowed to participate in the construction of the national crisis narrative, which may have resulted in a stewardship of the storyline and, in turn, winning a wider acceptance of the response narrative.

Heath's (2004) theory of crisis response introduces narrative theory into the realm of public relations and offers the audience a rational view of how stories converge from organizations, stakeholders, and other participating publics to co-create an acceptable, rational explanation of a crisis exigency. Heath's theoretical contributions aided this study in understanding the Chinese people's search for order in the SARS crisis and the central role of narratives in the government's communication response.

However, this study posits, at least in the context of the SARS crisis narrative, the individual's "search for order" (Heath, 2004, p. 167), within the evolving narrative, cannot be solely attributed to a "rational explanation of the event" (p. 168). Instead, individual acceptance of a crisis narrative may be attributed to both rational and emotional appeals of the co-created crisis narrative. The depiction of the Chinese medical professionals did not focus on providing a rational account of the crisis, but rather established an avenue for individuals to emotionally identify with the heroes.

The social context of a crisis event also plays an active role in an individual's search for order. Individuals may possess varying needs that are determined by the social context of the crisis exigency. In the SARS crisis, the Chinese people were confronted with competing narratives at the onset of the SARS epidemic, creating a social context that warranted a different approach. The government's initial failure to accurately represent the situation, which exacerbated the crisis exigency, placed them in a position where a rational and factually accurate narrative would not be as compelling as the competing narrative. The addition of the emotionally appealing narratives in the *People's Daily* served to fulfill the weakness of the rational appeal present in the existing social context. In such a social context, adding the emotional component strengthened the appeals of the crisis narrative. The addition of the emotional appeals and social context offers an extension of Heath's crisis response theory.

REFERENCES

An epic of life—memories of Dr. Liang Shikui. (2003, May 4). *People's Daily.* Retrieved April 24, 2004, from http://www.snweb.com/gb/people_daily/2003/05/04/a0504008.htm

Background stories of the Nurse's Diary. (2003, April 25). *People's Daily.* Retrieved April 24, 2004, from http://www.snweb.com/gb/people_daily/2003/04/25/e0425001.htm

Beijing school kids enjoy a spring outing. (2003, April10). *People's Daily.* Retrieved April 24, 2004, from http://www.snweb.com/gb/people_daily/2003/04/10/d0410006.htm

Benoit, W. L. (1995a). *Accounts, excuses, and apologies: A theory of image restoration strategies.* Albany, NY: SUNY Press.

Benoit, W. L. (1995b). Sears's repair of its auto service image: Image restoration discourse in the corporate sector. *Communication Studies, 46*, 89-105.

Bezlova, A. (April 27, 2003). Beijing tightens restrictions to isolate SARS [Electronic version]. *USAToday*. Retrieved April 10, 2004, from http://www.usatoday.com/news/health/2003-04-27-sars-beijing_x.htmat

Birthday song at Xiaotangshan SARS Hospital. (2003, May 7). *People's Daily.* Retrieved April 24, 2004, from http://www.snweb.com/gb/people_daily/2003/05/07/k0507006.htm

Building the line of defense against SARS—Frontline medical workers in Beijing visited. (2003, April 23). *People's Daily.* Retrieved April 24, 2004, from http://www.snweb.com/gb/people_daily/2003/04/23/b0423004.htm

China was being responsive and timely on SARS epidemic. (2003, April 4). *People's Daily.* Retrieved April 10, 2004, from http://www.snweb.com/gb/people_daily/2003/04/04/b0404003.htm

CNN. (April 24, 2003). *Timeline: SARS outbreak.* Retrieved April 10, 2004, from http://www.cnn.com/2003/HEALTH/04/24/timeline.sars/

Cowden, K., & Sellnow, T. L. (2002). Issues advertising as crisis communication: Northwest Airlines' use of image restoration strategies during the 1998 pilot's strike. *Journal of Business Communication, 39,* 193–219.

Fighting against SARS with her flesh and will—memories of Dr. Li Xiaohong. (2003, May 4). *People's Daily.* Retrieved April 24, 2004, from http://www.snweb.com/gb/people_daily/2003/05/04/d0504001.htm

From Guangdong—how are the SARS data collected and is it reliable? (2003, April 10). *People's Daily.* Retrieved April 24, 2004, from http://www.snweb.com/gb/people_daily/2003/04/12/b0412001.htm

Gods of Protection for patients—on the stories of Interior Mongolia Hospital. (2003, May 7). *People's Daily.* Retrieved April 24, 2004, from http://www.snweb.com/gb/people_daily/2003/05/07/b0507004.htm

Guangdong convalesced patients visited and interviewed. (2003, April 9). *People's Daily.* Retrieved April 24, 2004, from http://www.snweb.com/gb/people_daily/2003/04/09/e0409001.htm

Guangdong hospital visited—SARS is curable and controllable. (2003, April 10). *People's Daily.* Retrieved April 24, 2004, from http://www.snweb.com/gb/people_daily/2003/04/10/b0410007.htm

Hearit, K. M. (1995a). From "We didn't do it" to "It's not our fault": The use of apologia in public relations crises. In W. N. Elwood (Ed.), *Public relations inquiry as rhetorical criticism* (pp. 117–134). Westport, CT: Praeger.

Hearit, K. M. (1995b). "Mistakes were made": Organizations, apologia, and crises of social legitimacy. *Communication Studies, 46,* 1–17.

Heath, R. L. (2004). Telling a story: A narrative approach to communication during a crisis. In D. P. Millar & R. L. Heath, (Eds.), *Responding to crisis: A rhetorical approach to crisis communication* (pp. 167–187).

Heath, R. L., & Millar, D. P. (2004). A rhetorical approach to crisis communication: Management, communication processes, and strategic responses. In D. P. Millar & R. L. Heath (Eds.), *Responding to crisis: A rhetorical approach to crisis communication* (pp. 1–18).

Hong Kong 138 SARS patients convalesced. (2003, April 9). *People's Daily.* Retrieved April 24, 2004, from http://www.snweb.com/gb/people_daily/2003/04/09/b0409007.htm.

Kauffman, J. (2001). A successful failure: NASA's crisis communications regarding Apollo 13. *Public Relations Review, 27,* 437–448.

PCOnline. (December 15, 2003). *Reading the top 10 Careers of 2003 in China.* Accessed October 12, 2004, from http://arch.pconline.com.cn/pcjob/rs/zc/0312/253379.html

People's Daily Online. (n.d.). Introduction to *People's Daily*. Retrieved April 30, 2004, from http://english.people.com.cn/other/about.shtml

SARS patient recovered, with photo. (2003, April 10). *People's Daily.* Retrieved April 24, 2004, from http://www.people.com.cn/GB/paper464/8919/832151.html.

Seeger, M. W., Sellnow, T. L., & Ulmer, R. R. (2003). *Communication and organizational crisis.* Westport, CT: Praeger.

Sellnow, T. L., Ulmer, R. R., & Snider, M. (1998). The compatibility of corrective action in organizational crisis communication. *Communication Quarterly, 46*, 60–75.

The epical story written with his life—on the life stories of Dr. Deng Lianxian. (2003, May 4). *People's Daily.* Retrieved April 24, 2004, from http://www.snweb.com/gb/people_daily/2003/04/26/b0426007.htm

Time. (2003, April 8). *Beijing's SARS attack: Doctor and party member insists there are many more cases than officials will admit.* Accessed on October 31, 2004, from http://www.time.com/time/asia/news/article/0,9754,441615,00.html

United we fear nothing—on Beijing Tongren hospital. (2003, May 2). *People's Daily.* Retrieved April 24, 2004, from http://www.people.com.cn/GB/paper464/9083/845559.html

Venette, S. J., Sellnow, T. L., & Lang, P. A. (2003). Metanarration's role in restructuring perceptions of crisis: NHTSA's failure in the Ford–Firestone crisis. *The Journal of Business Communication, 40*(3), 219–236.

World Health Organization. (April 11, 2004). Severe Acute Respiratory Syndrome —*Press briefing*. Accessed October 31, 2004, from http://www.who.int/csr/sars/ Press_2003_04_11/en/

Zhang, X. (2004, from November 30). SARS crisis performance of the media's news reforms. Accessed on December 31, 2004, from http://academic.mediachina.net/xsjd_view.jsp?id=1854

8

NARRATIVES OF CRISIS PLANNING AND INFECTIOUS DISEASE

A Case Study of SARS

Robert L. Heath

Feng Li

Shannon A. Bowen

Jaesub Lee

Without careful planning and execution, an epidemic can spread quickly, killing or otherwise unnecessarily harming and alarming hundreds, thousands, and even millions. SARS had that potential because the appropriate authorities in China were caught up in a dysfunctional cultural narrative that blunted the response protocol. This chapter applies narrative theory to argue that organizational life is narrative enactment. When the people involved do not know the appropriate narrative for crisis prevention and response (or are unwilling to enact it), the crisis emerges as a risk manifested. Properly enacted, the narrative brings control to the risk as quickly as possible. If not enacted properly, a crisis emerges with all of the characteristics of uncertainty and ambiguity. The news media feed on this narrative, often making matters worse. Such occurred until enactors from other nations exerted appropriate control and brought the SARS epidemic to a fitting close.

Chaos theorists feature the following narrative: The single flap of one butterfly's wings in Brazil can set off a series of weather events that seriously damage part of the United States (McKie, 2001). An amazing replica of that metaphor in reality set off the narrative of the Severe Acute Respiratory Syndrome (SARS) epidemic (2002–2003). Interspecies transmission of a virus led to crisis proportions in many public health networks, countries, and organizations (e.g., Nicholas, 2003). This case is worth close study for lessons learned.

During this epidemic, competing narratives emerged and for a while were at odds with one another. The narrative emerging in China led to the incomplete enactment of the broader, more global one that ultimately dominated organized global public health response. The narrative that emerged from China, starting at least on November 16, 2002, lacked the explanations and qualities needed for a responsive and responsible public health narrative. Such a narrative was ultimately required to enact appropriate protocols developing through public health agencies such as the World Health Organization (who) and the Centers for Disease Control (CDC).

Because of the uncertainty associated with crisis preparation and response, requisite variety increases the likelihood that the needed information will be gathered and analyzed so that appropriate messages will be set into play. Rather than totally random behavior, crisis preparation and response needs focus (Holder, 2004; Weick, 1988). A crisis tends to be a time of uncertainty and ambiguity (Sellnow & Ulmer, 2004). Responses to the rhetorical problem that emerges from (or leads to) the crisis are required to bring control through communication processes that generate useful meaning. Through act, react, and interact, messages are set forth, examined, and reexamined (Holder, 2004). Through debate, they are incorporated into a working narrative that can consolidate investigation of and response to a crisis event, leading to sound conclusions and appropriate response protocols. As messages suffice or fail, they are incorporated into a working narrative that can coordinate further efforts on the crisis, such as in the case of the SARS epidemic. In one sense, the efficacy of public health narratives depends on the ability for the narratives to protect the public health maximally in conjunction with known science. Conflicting or incomplete narratives are likely to *lead to* and complicate rather than prevent or truncate crisis.

To examine this logic, this chapter features narrative crisis response as a case study of the impact and handling of SARS, which emerged onto the international public health scene in 2002–2003. This potentially fatal disease affected people and then organizations, first in China and following quickly around the globe. The case began with the basic rhetorical problem and scientific challenges of defining, understanding, and reporting the disease, its symptoms, contagion, and preventive or corrective measures. Failure to adequately communicate about the epidemic at the regional and national levels in China prevented broader and more functional public health protocols from working optimally in emergency response (Fritsch, Pottinger, & Chang, 2003). This was a volcanic epidemic as opposed to a stalagmite crisis that builds up one drip at a time until it reaches the boiling point of crisis. Failures often teach us more than successes in such unfortunate situations.

In the SARS epidemic, events occurred with such speed that highly refined response protocols were needed from the outset. As Heath (2004) asked, "Does the narrative enacted by the organization responsible for the crisis demonstrate that the organization has not lost control or that its efforts will restore or achieve control; is the resolution of the story satisfactory to the key publics?" (p. 177). Credible crisis response narratives must hold up under lay and expert scrutiny; facts and plots need fidelity, probability, and coherence (Fisher, 1987).

This chapter argues that, because of systemic, rhetorical, and ethical flaws, the narrative that originally arose during the SARS outbreak was incomplete, slow to develop, and framed defensively to protect certain interests—to the potential detriment of other interests. Poor management of this crisis led the risk bearers (lay and professionals in public health; Palmlund, 1992) to be improperly warned of the potential impact on them, and they did not have appropriate intervention to protect their interests. Specialists in public health who should have been risk intervenors were actually risk creators. The ethical conundrum resulting from these missteps is one of huge proportions, with life and death stakes. A power-serving variance in the basic principle of the value of human life appears to underlie such judgments in the SARS case.

NARRATIVE FRAMES

Every crisis exhibits unique as well as generic form, content, process, and relationship challenges. By type, any crisis has its peculiar set of narrators, auditors, plot, moral (theme), characters, location (setting or scene), and relationships (Heath, 1994). The logic of a narrative approach to crisis began as academics adopted and refined the work of Walter Fisher (1984, 1985, 1987), who proposed that humans be studied as *homo narrans*. Fisher argued that humans' knowledge and coordinated activity result from a complex of shared narratives. Human existence consists of many lived narratives that allow for and guide coordinated as well as conflicting preferences, choices, and behaviors.

Narrative theory reasons that humans not only tell stories, but also live them. Narratives define and specify personae, scripts, roles, locations, plots, thematic morals, and relationships. What any group of individuals does and says becomes meaningful to themselves and others because of how their narratives interpret and guide what they see, hear, and do. In this way, narratives give order to society. They form definitions, relay belief systems and values, express expectations, and normatively guide individual and collective choices. Some narratives lead to, grow out of, and resolve crisis. Others may frustrate that process. Therefore, a narrative approach to crises allows us to test and refine crisis theory and generate best practices for crisis planning, prevention, and response.

Depending on definitions, each crisis falls somewhere along a continuum featuring bad publicity, planned emergency response, and chronic and even fatal crisis. As Heath (1997) reasoned, crises vary in magnitude ranging from (applying a medical metaphor) bed rest to medication, chronic conditions, and death. Response experts define a *crisis* as a set of threatening challenges that normal emergency response protocols and capabilities cannot successfully or easily manage.

Narratives serve as the basis for rhetorical identification (Burke, 1969a, 1969b). Shared narratives allow for or facilitate coordinated activities to help people enact society as the collective management of risk (Douglas, 1982, 1992). The challenge to manage risks motivates any society or segment of a society to prepare for, respond to, and learn from crises. Some people are affected, whereas others are not. One universal use of narrative is to share warning signs and response protocols appropriate to risks encountered. The science and protocols of risk detection and crisis preparation help people—lay and expert—to detect the potential for a risk to manifest itself. Crises vary in terms of magnitude of severity. Likewise, as narratives, crises exhibit varying degrees of drama.

This SARS case study serves as an example of a crisis that—if properly managed—could have been remedied quickly with medical treatment. However, SARS spread and became a chronic crisis because inappropriate narratives dominated the thinking of various officials and the general public. Mismanagement of the crisis can erode and even destroy an individual's or organization's integrity and right to operate in the public interest.

The initial narrative failed at least in part due to cultural narratives that were inherently dysfunctional and competing with other, ostensibly more dominant narratives of public health response. Persons who shoulder the responsibility of professionally engaging in risk detection, management, and mitigation require skills, training, as well as investigative and communication protocols. These established protocols serve to bring forward useful information and follow planned guidelines in a timely fashion to mitigate the impact of the risk on risk bearers. Risk managers then could exert the appropriate control needed to avert and recover from the crisis. Problems in managing the SARS epidemic arose from conflicting cultures. Cultural values undoubtedly influence the belief systems that are used in decision making. Sriramesh, Grunig, and Buffington (1992) observed that culture influences interpretations of fact, ethical responsibility, rhetorical problems, and strategic responses (including message design and delivery). Such logics are especially relevant to crisis preparation, response, and postcrisis lessons learned (Bechler, 2004; Holder, 2004; Marra, 2004; Stacks, 2004; Turner, 1976).

Culture not only predicts what will be said during a crisis, but also how capable an organization is of signal detection, a critical stage of crisis prevention and response (Coombs, 1999; Pauchant & Mitroff, 1992). If a crisis cannot be observed early, as it is emerging, then communication and procedural responses are likely to be difficult at least and dangerously inadequate at the extreme. Narratives of public health protocols must contain role assignments and scripts for the early detection of crisis onset. Quick and effective response is vital in the case of public health crises. Narratives at play will predict whether the response is quick and effective. In terms of SARS, Feldman et al. (2004) noted that flows of information eventually produced a coherent story that emerged as conflicting story lines collided with one another and eventually refined each other. Narratives reflecting Chinese culture were in conflict with dominant cultures of Canada, the United States, and the World Health Organization (WHO); eventually their narrative prevailed, but not without clash, conflict, and concern.

FRAMING CRISIS COMMUNICATION IN THE PUBLIC HEALTH ARENA

By definition, a crisis is an event or issue trajectory that threatens the ability of one or more organizations to exert control in a manner that is responsible to the public interest or, even more specifically, to the interest of one or more publics. Does its reaction and recovery meet or exceed the needs of key publics? The event or issue is associated with outcomes of varying degrees of uncertainty and magnitude. Crisis is characterized by uncertainty or instability in regard to the performance of an organization, as well as its ability and willingness to work on behalf of the interests of its publics. What harm has or might be done, to what degree, to whom, and for what reasons are standard narrative elements of the rhetorical problem that arises during and after a crisis.

Crisis affects normal communication processes, best practices, and strategic messaging. Processes become intense and focused on getting information, interpreting that information, and providing it to people who are interested in the event. Messages offer facts, interpretation of facts, evaluations, and policy or procedures. Decision makers hold the grave responsibility of managing the interests of others (Heath & Millar, 2004). Barton's (1993) exacting definition of crisis stressed that crisis results when an event or issue strains an organization's ability to cope using its ordinary procedures.

Perhaps few cases are more relevant to this view of crisis than are those that arise from the public health arena. Potential impact can be severe, from injury to death and epidemic. Uncertainty can be gripping, even debilitating. Information vacuums can fill quickly with misinformation, myth, mystery, and even disinformation, compromising the interests of risk bearers. These types of crises have perhaps the highest stakes: life and death for many. Public health and even future lives are at stake.

By this definition, the 2002–2003 SARS epidemic posed one of the most compelling recent public health crises, with the health of the people in China, Hong Kong, Singapore, Vietnam, Taiwan, and Canada at heightened risk. Also at risk were the protocols traditionally implemented at the outbreak of a disease with serious public health consequences. Any epidemic can be handled poorly or well by public health officials. If the protocol used is inappropriate or mishandled, the skill and commitment of those entrusted to enact it may be called into question. The crisis can be created or augmented by uncertainty about the efficacy of the protocol—either its design or operation. The SARS crisis increased in magnitude when little was known about its nature and origin and the public health consequences. To make matters worse, China's public health authorities were not vigilant and probably hid or at least obscured the truth for economic reasons. They did not know or simply avoided implementing public health protocols relevant to the type and seriousness of the disease (Fritsch et al., 2003).

In the early stages of what seemed to be a routine influenza epidemic, officials in China may be excused of responsibility because they did not understand or appreciate the seriousness of the disease; they often merely sent contagious individuals home from work or to a clinic as victims of the flu. Under the circumstances, officials were reactive because they did not proactively assume that the disease might be extraordinary. In fact, in what was perhaps the first ethical lapse, they wanted to downplay this prospect or the reality as it emerged. Thus, that reactive narrative helped spread the disease and terrified citizens and public health officials. More contact with contagious persons led to more disease and so forth in a cascade. It becomes clear that officials needed either to develop a better narrative protocol or use the existing one more ethically and skillfully to bring control over the crisis.

One example of the conflicting narratives that played out centered on domestic animals. In April and May 2003, police, government authorities, local neighborhood committees, and individuals in China began to slaughter thousands of pets, including dogs and cats, they feared were carrying and transmitting SARS. Such actions probably had no positive effect on the outbreak, but it did have a devastating impact on pets and the pet food and pet care industry, leading Edelman Public Relations' Beijing office to bring third-party testimonials into the fray. The agency seized an opportunity in a culture not prone to respect animals or highly regard the opinions of veterinarians. The opportunity came in May 2003, when *People's Daily* quoted a virologist, Professor Shen Rongxian, from the prestigious Academy of Sciences saying that there was no evidence that pets transmitted the coronavirus that caused SARS. The agency moved quickly and set up a meeting between Professor Rongxian, the veterinarian association, and a Chinese government institute responsible for the study of disease transmission through animals. Within a week, the number of pets dumped into the street or slaughtered began to decline. By the end of May 2003, Hong Kong researchers identified the civet cat, a wild rodent-like creature, as the primary carrier of the coronavirus that caused SARS. This further eased the focus on domestic animals (Nicholas, 2003).

For centuries, public health officials have refined the narrative protocols necessary to serve the public interest in the event of a contagious disease outbreak. The thematic logic of this narrative begins with understanding the disease—its nature, symptoms, means for transmission, treatment, and protocol for containment. Once these facets of a disease are understood, protocols are developed and communicated to professionals as well as lay publics with the intent of abating the spread of the disease and mitigating its effects. The thematic moral logic begins with knowing the disease and its modes of transmission, developing or applying protocols, and communicating with the interested and relevant parties. What is important is the difference between known and unknown diseases. If a disease is known and an outbreak occurs, it is likely that protocols are in place to deal with this disease, constituting a routine narrative. But if the disease is new or not well known, the narrative develops in quite different ways akin to chaos theory.

Set against these criteria, SARS was a crisis the moment it began to emerge onto the world public health scene as an unknown yet deadly disease. Mystery shrouded its nature and the appropriate response to contain it. One primary condition for avoiding a crisis is for responsible people and organizations to have and implement well-tailored narrative protocols to discover, investigate, report, and respond to medical crises. This narrative requirement assumes that all of the responsible parties immediately implement appropriate protocols. That protocol requires not only openness, but also an absence of denial and dissembling. Narrative defines not only the appropriate medical professional responses, but also constitutes the message development and dissemination that invariably ranges from that of one clear, professional, and responsive voice to a Babel of confused and contradictory voices that may become increasingly alarmed and conflicting. The voices may add to and even cause the crisis if the Babel is sufficient and the narrative is confusing and insufficient to meet the medical requirement. Experts need to understand and address the various rhetorical problems that contextually are relevant to communicating appropriately with a variety of publics. The goal is to protect and promote public safety in a responsible, ethical manner.

FRAMING CRISIS COMMUNICATION AS NARRATIVE

Narratives operate according to well-established elements: narrators, auditors, plot, moral (theme), characters, location (setting or scene), and relationships (Heath, 1994). By protocol, central narrators in public health cases are trained members of the public health community who work in conjunction with media networks and interpersonal contacts in the medical profession, along with patient contact. In the systemic communication process, interpersonal contact is invaluable. People talk to one another about diseases. The resulting narratives frame and constitute messages. The stories guide and result from actions. They serve to frame and structure meaning and shape understanding and agreement. Narrators are those persons who tell the stories as well as serve as the directors of the enacted narratives. Narrators are networks of public health officials, other government officials, members of various industries, reporters, friends, and family.

Auditors are people who do or are expected to listen to the story. They include publics the narrator health professionals believe need to know about the disease and take appropriate action. These auditors are likely to be different if they are health professionals as opposed to lay publics. They may be hospital personnel wanting information that allows them to enact specific narrative protocols. They may be a risk-bearing public, media personnel, and persons in various industries and relevant organizations that can be or are affected by the disease. For instance, health professionals and lay publics may be told various symptoms that make diseases recognizable. Health professionals may have learned these symptoms and corrective measures as part of their specialized training. In the advent of a new disease, they need information of this kind—along with specialized information, such as diagnostic tests, which may not be relevant to the lay publics who lack the instruments and training to acquire and interpret the data. In such cases, the rhetorical problem is: How can this disease be detected and treated? The general public and various other publics need some of this information, but at a lower level of technical specificity.

One of the outcomes of the SARS epidemic was the ability to recognize and categorize the disease by analyzing its DNA sequence. Likewise, researchers discovered the likelihood that it originated from interspecies human contact with wild animals. The cultures of parts of Chinese society prevented both clear understanding of this contact and proper preventive behavior because sources of income and food were at risk—accurate information might lead to public health protocols against owning and slaughtering such animals. Culture probably prevented appropriate early response narrative interaction between narrators and auditors of various kinds. Flows of information are expected to be expeditious, but can, for reasons such as these, be substantially dysfunctional. Reasoning based on dysfunctional information and logic is rarely sound.

By design, the plot should develop in an orderly narrative based on national and international public health protocols. These are communicated around the world, especially through the efforts of the WHO. The ideal plot features an early discovery of the disease, a rapid understanding of its nature and degree of threat, insights into its transmission, and expeditious responses. If normal protocols are appropriate, the narrative calls for them. If new ones are needed, the narrative must accommodate that prospect. For example, resistant diseases call for narratives of persistence. The ideal plot, then, is disease relevant. Simply, the plot for a contagious disease is different than for one that is not transmitted, genetic, or difficult to acquire. Understanding and responding, for this reason, can occur according to known and understood public health plots. SARS achieved a crisis level by the enactment of dysfunctional plots that frustrated rather than facilitated effective response.

The thematic moral of public health narratives revolves around the proficiency of the plot (system, including multiple, interlocking protocols), characters, and the specific disease. The central theme of this narrative features efficiency of gathering, sharing, and analyzing information. Facts are disseminated according to protocols to people on an as-needed basis. If the system works correctly—and the disease is responsive to our understanding, it is countered and abated with a minimal loss of life or other damage. Some diseases, such as AIDS, pose extremely complex narratives. In the case of SARS, the theme—lesson learned—was that appropriate protocols are known—even if they need continual refinement—but can be frustrated when the prototypical plot is not followed because an alternative, dysfunctional one is followed. At the macrolevel, the preferred theme is to get from "once upon a time" to "happily ever after" as quickly and ethically as possible, lessening damage to people and associated interests.

Relevant to this study is the thematic moral featured in a *Communication World* best practices study of SARS. DuHamel (2003) noted the extraordinary crisis response required for the SARS epidemic. This situation did not fit traditional crisis management strategies where organizations apologize for an error or problem and then correct it. The situation required the health care profession to do its best to protect its staff and the people who look to them for help. A crisis offers opportunity for communication professionals to play a key role and show how they can navigate the tricky waters of an uncertain situation; ethical and effective communication makes the difference between success and failure. DuHamel reasoned that,"No matter how bad the situation gets, every story has a beginning, a middle and an end" (p. 48).

Narrative characters are the roles taken by individuals who enact a plot. In public health cases, a lot of the roles are defined either by public health protocols or the ancillary effects of the disease on persons outside of that profession, but whose professional and personal lives are affected. In the SARS case, public health organizations suffered varying degrees of crisis as they implemented protocols, with varying degrees of expertise and willfulness. Protocols identify an epidemic, define its nature, develop appropriate response protocols to mitigate its spread, implement controls over the contagion, and communicate within the public health community and at-risk publics.

People were affected who are responsible for economic development and safety of people in schools and workplaces. Early detection and understanding of SARS was a challenge to various characters—ranging from cooperative to hostile—in government and public health arenas. Once the disease became better understood and people in other professions beyond public health recognized its implications, those characters took on new life and became part of the total narrative. For instance, if the disease were only occurring in China, the travel, tourism, and public health officials of a country such as Canada would have no concern. Once SARS victims (characters) appeared in Canada, those characters began to experience crisis and became players in the total crisis. Individuals—the general public experiencing its own sense of vulnerability without much public health assistance—took extraordinary measures such as wearing protective masks to reduce the likelihood of contagion. Uncertainty was high. Knowledge was low. Combined, these factors produced a high level of anxiety. Self-protection became the theme of characters taking responsibility for public health.

Location is not a trivial element of public health narratives. Location can be physical or more symbolic. A remote part of Southern China, Fushan City, Quangdon Province, with its unique culture seems to have been the point of origin for this disease from which it spread to other parts of China and to the world (Zhao et al., 2003). Had it remained local, the crisis would have been different. Because of the silence, deceit, and ignorance relevant to the disease's nature and origin, it spread quickly, relentlessly, and ruthlessly. It crossed borders. As it spread, it engaged different characters. Once that occurred, the narrative took on many permutations, including one that suggested that the public health system was strained and even constrained in its efforts to understand the disease and develop, report, and implement protocols. Public health systems and protocols seemed dysfunctional. Calls in many countries demanded changes in global public health protocols. One arose in the U.S. House of Representatives where legislators argued that Taiwan needed to be in the WHO because it would not make the mistakes of judgment that occurred in China (La Rocque, 2004; see also Taiwan Security Research, http://taiwansecurity.org/TSR-SARS.htm).

Culture affects narratives as well as being expressed in them. Culture has a direct impact on organizational behavior, professional performance, and society. Although China has experienced substantial reform in social structure and economic systems, the traditional values of its culture, such as collectivism or Confucianism, still play a dominant role in the evolution of this transition. Confucianism is a deeply rooted foundation of every aspect of social structure, family values, social relationships, and organizational management. Confucianism focuses on hierarchical relationships of a society, wherein members are kept in order. Society is a hierarchical pyramid of roles that people enact in relationship to other people in complementary roles. Roles define relationships, which in turn shape roles. Subordination and superiority are natural and proper in hierarchical social relationships. These principles provide the basis for Chinese organizational bureaucracy, but are inadequate in emergency preparation and expeditious response.

The SARS narrative was clouded in mystery and high drama for perhaps longer than necessary given what was ultimately understood about the disease in light of the protocols that could contain its spread and minimize its damage. One of the most dramatic elements was the fact that lay people and medical professionals appeared in public wearing surgical masks. This character and plot added drama to the narrative theme that was evolving. The case reached crisis because government officials acted out of fear that information about the disease could be harmful to their interests, specifically economic development.

SARS took a public voice on February 21, 2003, in Hong Kong when a doctor from southern China attended a family wedding. Reportedly, he spread the disease to other members of the wedding party and hotel employees. From this epicenter, infection occurred quickly as he returned to the hospital and members of the wedding party dispersed through the Hong Kong airport. Infected persons spread the disease to Singapore, Hanoi, and Toronto. Crisis occurred when the doctor died. Over the next weeks, 700 more people died from the disease. One character, the WHO, began to understand SARS' transmission and protocols for abating its spread and deadly effects (www.cbc.ca/disclosure/archives; see also Fritsch et al., 2003).

This epidemic of contagion and fear occurred despite public health officials' ability to identify the strain by examining its DNA sequence information. This investigation could be done in 24 hours. Through DNA, scientists were able to trace the origins of the disease to wild animals. To do so, however, they had to cope with local resistance, cultural denial, and authorities who were willing to look at this problem as anything *other* than one of public health.

By 2004, the DNA nature of the disease was well documented as a coronavirus. The strain can be fatal, especially for individuals who have weakened immune systems. It can cause severe and acute reactions in animals, including cats, dogs, pigs, and birds. The virus is conveyed through personal contact. Several routine laboratory tests can be used to identify its presence.

By 2004, the U.S. Center's for Disease Control (CDC) listed guidelines and information relevant to various publics: clinicians/health care settings, health departments, laboratories, legal communities, patients and their close contacts, schools and workplaces, and travelers. The CDC also offered travel advisories (CDC.gov).

THE SARS NARRATIVE: AN EVOLVING EPIDEMIC CLOUDED IN MYSTERY

However shrouded in mystery, this narrative began with "once upon a time" and struggled to end "happily ever after" because the early plot that emerged did not sustain itself under public scrutiny. The original narrative lacked fidelity, ethical validity, and probability; it also lacked internal consistency. Information was incomplete, slow to emerge, and framed defensively to deny or downplay the seriousness of the disease. Risk bearers were at more risk than they were told; the narrative could and did lead them to both underestimated and undue concern. Risk intervenors—public health authorities—were kept out of the narrative far longer than protocol prescribed or ethics warranted. The risk creator, or virus, was knowable given the status of medical diagnostic technologies, but needed to be examined and understood expeditiously. The root cause—exposure to wild animals as food sources—was confounded because culture and fear obscured the vigilance of the public health authorities. To reveal wild rodentlike creatures being used as food sources would be embarrassing to the economic ministry of China. Further complicating this mix was collectivist Chinese culture, which can be indifferent to individual rights and the protocols preferred by the WHO and CDC.

After SARS surfaced in China's Guangdong, its quick spread provoked fear, in part, because it seemed to be eluding the normal efforts of the public health community to recognize the problem and respond with well-defined protocols. That rural China was the point of origin for the SARS outbreak tells an incomplete story. In terms of crisis, not only did people in rural China suffer from the threats of the crisis, but so did various branches of the Chinese government. Those branches included the central authority, public health officials, and persons responsible for economic development and tourism and travel.

Almost as a harbinger of the outbreak of the disease, on November 23, 2002, WHO held a conference on influenza in Beijing. Although it was near the emergence of the disease at an opportune time and with the right public health professionals, WHO was not as effective as needed in the early stages of the outbreak (www.cbc.ca/disclosure/archives). Just days before, on November 16, the first known case of SARS was discovered. By February 11, 2003, the Chinese Ministry of Health reported that more than 300 cases of SARS had been detected and 5 deaths had occurred in that province, and Hong Kong also reported an outbreak (Fritsch et al., 2003).

On March 15, 2003, WHO confirmed the outbreak and reported that SARS was now global: Canada, Indonesia, Philippines, Singapore, Thailand, and Vietnam. WHO issued warnings for persons traveling in Southeast Asia. Later in March, the disease had spread to the UK, Spain, Germany, and Slovenia. Further travel restrictions were issued. On March 29, Carlo Urbani, the WHO official who first identified SARS, died of the disease (BBC News, 2004). Of the events leading up to March 2003, Meng (2003) concluded:

> By March 2003, the spreading of SARS in north China had become more than dangerous. Yet the responsible agencies in the government were still blocking the information to not only the general public but also the medical institutions. It caused the hospitals (to be) totally unprepared in terms of knowledge and materials. Up to early April, when the epidemic ran rampant in Beijing, an official of the Ministry of Health insisted in public that there were only 12 SARS patients and 3 SARS-related deaths. "The SARS in part of China is under control; it is safe to work, to live and to tour in China!" (n.p.)

April 2004 was a crucial month in expanding the SARS narrative. Investigative journalists began to press public health officials in China for details. Under pressure, officials no longer could restrict the story. In the midst of publicizing pleasant experiences by tourists to China, a competing narrative of serious and deadly health problems began to emerge. Even before April, stories had asserted that something was amiss, but confirming details were lacking because of denial by officials. One reason for the *cover-up* was the mystery of the disease. It seemed both normal and abnormal to public health officials. It seemed merely to be the flu, a severe form perhaps, but not different.

Local protocols were put into place, and little effort was expended to identify the nature of the virus. Part of the reason for the slow news attention was the location of the largest number of cases. They were occurring in regions with the highest economic stake—and where some of the highest population density exists. Economic incentives of promoting tourism and industrial growth collided with sound public health protocols. Local and national officials wanted the seeming epidemic to be found to be normal, not extraordinary. They were quite sensitive to the economic damage that could occur if the epidemic were discovered to be extraordinary—with increased likelihood of being fatal (Meng, 2003). Media reports and Internet accounts began to bring details to a wider, more global, and more proactive set of actors in the narrative. The original story told by Chinese officials lacked narrative fidelity and moral veracity. It could not sustain itself under the increasing information demands of the global public health microscope.

By April 2003, WHO discussed a coronavirus never seen in humans as the cause of SARS. Perusal of the Web site of the CDC indicates special categories of characters, information, and protocols relevant to each disease. This is the fully functional paradigm of public health, against which the response to SARS is measured and seen as inadequate. In April, China apologized for its slow response to the epidemic. The announcement alleged that government officials had covered up the extent to which the disease had spread—by both geography and number of victims.

By May 2003, China issued public health warnings and protocols. It forbade spitting in public places and restricted entertainment venues in Beijing. Ten thousand people were quarantined in Nanjing. Official sanctions escalated in severity. The government threatened to execute or jail for life anyone who broke one of the quarantine restrictions. Further travel restrictions were imposed.

SARS had emerged as an unknown threat. Three unknowns characterized the disease: What is its cause? How is it spread? How deadly is it? Protocols that could have been implemented to address these unknowns were frustrated by the Chinese officials in Guangdong Province, especially from November 2002 to January 2003. These crucial 3 months were lost or wasted in responding to the epidemic, and needless deaths resulted. Also the medical community and general public likely were unnecessarily alarmed.

By the time SARS was contained in 2003, it had infected approximately 8,100 people and accounted for almost 800 (774 according to CDC report) deaths worldwide: China—349 documented; Hong Kong—299; Taiwan—37; Singapore—33; and Vietnam—5. In late July 2003, the epidemic was under control, but new cases were reported by the Chinese Minister of Health as late as July 2004. CDC and WHO monitored the situation in China closely throughout 2004.

SARS was a mystery disease creating a crisis not only for the public health community, but also for the tourism industry and government officials working for economic development. Officials worked to suppress information about SARS to keep potential investors and workers from worrying that their financial and personal well-being would be harmed by the disease. Tourism dropped 70.5% in China, 86.8% in Hong Kong, and 70.7% in Singapore, according to CNN International.com update on August 9, 2004. For several days, air traffic ceased in Toronto as a means to stop or at least slow the spread of the disease. This crisis affected the entire travel and tourism sector.

Hong Kong's public health officials recognized the failings that had occurred early in the outbreak. Part of its response was to enact color-coded alerts as a means for communicating risks and preventative measures to its citizens. By May 2004, WHO had determined that the person-to-person network that had transmitted this disease had been broken. The level of alert was accordingly lowered. Even so, various protocols were advised for citizens who might be exposed. These included a no visiting regulation at isolation wards. Persons who were high in potential exposure were advised to wash their hands and wear masks as precautions to exposure and transmission. Assurances of continued vigilance were issued by public health officials (Hong Kong Department of Health, 2004).

In addition to the victims who suffered and died from the disease, many other people also bore risk. Government officials seemed, at least in part, immune to punishment because either they had done the best they could or they were protected by their roles in the narrative. At least one victim, however, in China died because of incompetence or lack of proper training. The latest outbreak of the disease occurred in July 2004 because of the failure of the Chinese Ministry of Health to follow proper protocols. Li Liming, the director of China's main disease control center, resigned because of an April outbreak in one of its laboratories. In July 2004, Hong Kong Health Secretary Yeoh Eng-kiong was forced to resign because of the sharp criticism that accused him of being inattentive to the appearance and spread of the disease. He also was found guilty of spreading false reports to the Hong Kong public (BBC News, 2004). He had to apologize to the public after experts revealed significant shortcomings in the city's health system. One finding recommended that information be exchanged and analyzed openly and quickly in an international forum of experts leading to effective contingency planning ("Hong Kong Health Chief Apologizes," 2004).

Eventually one clear and coherent narrative emerged triumphant out of conflicting narratives. The flow of the disease prompted various public health organizations and relevant media to dig for the true story. Not only had different accounts conflicted, but various interests promoted or framed their preferred narratives. Feldman et al. (2004) discovered that some business interests wanted travel advisories issued by WHO lifted, physicians urged caution, and politicians battled for their own interpretations. Not only did different stories play because of conflicting motives, but also—especially in a global epidemic—it is likely that distortion and conflicting interpretation is a distinct problem.

Narratives must assist and empower public health personnel. Hall et al. (2003; see also Maunder et al., 2003) observed how the SARS response clearly was not business as usual. This plot was dysfunctional in the face of this epidemic, especially for public health workers. At times, this narrative required extraordinary patient screening, visitor control, use of cumbersome protective gear while attending patients, patient quarantine, and heightened security. As actors, the stress culminated on nurses who experienced fear of contamination, anxiety, frustration, isolation and a sense of loneliness due to work overload, and uncertainty in the work environment (e.g., rapidly changing infection-control directives). The normal plot of nursing encourages interaction to share protocols and provide support. This emergency response protocol in fact required the opposite, leaving nurses without their normal supportive narrative. Public stigma extended to nurses as well as their family members. Children of nurses were barred from school trips, and their families were shunned by neighbors. Husbands of nurses were sent home from work; nurses were refused rides by taxi drivers. Single-parent nurses were unable to secure child care. Contrasted against this stigma, thematic moral characterizations in the media featured them as heroes, soldiers, and superhumans who willingly put themselves in danger to sacrifice their lives for the sake of others. Nurses were portrayed as altruistic professionals who showed skill, courage, expertise, compassion, and dedication to public health care (Yeoh et al., 2003).

Competing and conflicting stories also occurred elsewhere. When the WHO announced its travel advisory against the city of Toronto, the city took the defensive denial approach. The mayor said that he had no idea why an advisory has been placed against the city and rhetorically pondered where those making the charges are based. Rather than appearing in charge of the situation, Toronto officials went out of their way to tell the world how utterly surprised they were that the WHO would issue the advisory (Piovesan, 2003). Canadian officials' responses ultimately added to the confusion among narratives explaining the severity of SARS.

ETHICAL IMPLICATIONS: THE CHALLENGE OF MEETING COLLECTIVE INTERESTS

Crises, both at the detection stage and the response that occurs, pose ethical challenges. Perhaps moral responsibility is never more crucial than in the case of public health. Life or death for hundreds or even thousands hangs in the balance of public health decisions, making these situations a natural locus for ethical decision making (Bowen, 2002). Public health organizations develop detection and response protocols that are covertly or overtly sensitive to ethical paradoxes. Ethical issues are associated with crisis and are the principal factor in ethics scandals (Seeger, Sellnow, & Ulmer, 2003). Seeger et al. (2003) described five ethical perspectives that are particularly salient to crisis. The challenge and, therefore, the answer to these ethical challenges focuses on issues of organizational responsibility and responsiveness, ethical standards enacted by organization leaders, humanistic values specific to the ethics of care, issues of organizational legitimacy, and truthfulness and other significant choices that are central to the nature of crisis detection and response.

- Standards of organizational responsibility broadly refer to the fact that individuals and groups have obligations to others and to larger ethical and moral codes, standards, and traditions. Perhaps especially true of public health organizations, the fundamental rationale for any organization's existence is the value it adds to society (Johannesen, 1996).

The level of moral responsibility for an issue is determined by the amount of control the person or organization exercises over the decision. Moral responsibility requires that a person both knowingly and willingly engages in the action. Therefore, lower level Chinese officials and health care workers can be excused of the cover-up associated with the early phases of the SARS virus. However, decision makers high in the Chinese government cannot be excused because the inadequate and inaccurate information regarding SARS was their causal responsibility. Misinformation and disinformation resulted from a conscious choice by Chinese officials to obscure the true origin of the disease.

- Public health organizations by mission and design must create strategic relationships that enable them, as well as other organizations in the crisis, to respond effectively and quickly (Seeger et al., 2003).

In the SARS case, external relationships were not maintained or respected in the initial response of the Chinese government. Providing candid and accurate information as it was known would have been helpful to the WHO, CDC, and other health agencies. Unfortunately, the Chinese government had little relationship with these organizations and preferred to attempt to handle the SARS problem alone, within its borders. When the disease spread to other countries, the outcry of the scientific and health communities became such that the Chinese government was then forced to respond, rather than doing so out of an ethical obligation to world health.

- A humanistic orientation requires that public health organizations be sensitive to the harms that may be caused by their operations. Harms can result from decisions and acts of omission as well as commission (Seeger et al., 2003).

The harms resulting from the SARS epidemic were numerous. The harm was exacerbated in China by the reluctance of the Ministry of Health to disclose accurate and timely information regarding the source of the disease. These government officials judged their own interest as higher priority than that of risk bearers and the ill and did harm by omitting the true information known about the genesis of SARS. This type of decision making is unethical because it does not respect the publics whose health is at risk and it is the responsibility of the Ministry of Health. The decision also fails to meet other ethical tests (Bowen, 2004, 2005) because it is based solely on prudential self-interest as a decision-making guide. Using self-interest alone to guide ethical decisions is dysfunctional because such an approach negates the objective rationality required to effectively analyze decision alternatives.

- Legitimacy is conferred on the organization by key publics. To remain legitimate, organizational leaders must conduct themselves in ways that are judged by those publics as ethical (Seeger et al., 2003).

The Ministry of Health, by putting its own interests ahead of that of risk-bearing publics, lost legitimacy and credibility in not only their eyes, but also in the eyes of many other nations. The Chinese government lost legitimacy by being dishonest, not forthcoming with information, and placing its own interests ahead of Chinese citizens and other risk bearers around the globe. The result was an increase in the distrust of Chinese officials by Chinese citizens and by others in the world public health arena.

- Truthfulness and honesty, versus deception and lying, pose complex moral issues prior to and during a crisis.

Legitimate reasons for secrecy exist, such as patent and corporate secrets. However, if an act must be conducted in secrecy, ethicists (Bok, 1983; Kant, 1797/1994) concluded that concealment might indicate a moral problem. This is the case with the Chinese government's decision to conceal the source of SARS. The embarrassment caused by rural citizens foraging food sources from the wild civet cats carrying SARS was judged to be reason for withholding important information from the worldwide public health community. This decision was an error in judgment that is clearly unethical when viewed in terms of the potential world health impact of the

SARS epidemic. Government officials in Toronto also bear moral responsibility for not being honest about the extent of the danger posed by SARS.

LESSONS LEARNED FOR PUBLIC HEALTH AGENCIES: A NARRATIVE REFINED

Society operates according to various narratives. One that has become increasingly refined over the years offers protocols, systems, and messages relevant to public health. Without doubt, the early days of the emergence of SARS revealed how flawed various agencies were in this regard. In this case, both ethical decision-making systems and response protocols were flawed. Agencies were experiencing but denying crisis, as were citizens and various other organizations. The crises experienced by citizens and other organizations were magnified because key public health agencies failed to enact the narrative of responsible professional public health management. Narratives can either help resolve crisis situations or complicate them: In the SARS case, the epidemic was exacerbated by dysfunctional narratives.

One telling critique was performed by the public health system of Ontario, Canada. Campbell (2004) wrote for the Commission when he applauded the performance of individuals in the public health system while indicting the system:

> Despite the extraordinary efforts of many dedicated individuals and the strength of many local public health units, the overall system proved to be woefully inadequate. SARS shows Ontario's central public health system to be unprepared, fragmented, poorly led, uncoordinated, inadequately resourced, professionally impoverished, and generally incapable of discharging its mandate. (p. 118)

Part of the Commission's critique pointed to efforts to stifle or distort facts: "Any attempt to manage the news by stifling important sources of information will not only fail but will also lead to a loss of public confidence and a feeling among the public that they are not getting the straight goods or the whole story" (p. 119; see also Hearne, Hamburg, & Segal, 2004). To that end, an observation by EUROPA (2004) is illustrative:

> SARS has demonstrated the value of acting globally and in a coherent manner. The establishment of the European Centre for Disease Prevention and Control that has been proposed by the Commission in July 2003 would provide additional capacity to response more effectively and promptly to communicable disease threats.

Performance is the plot, and public interest is the thematic moral of this narrative.

Lessons learned are a vital part of any crisis and are especially important in this case to various public health agencies. The Health, Welfare and Food Bureau of Hong Kong posted on its Web site a Highlight of Government Anti-SARS Measures document in June 2004. The document recommended changes that were the result of a SARS Expert Committee. Hong Kong also announced an addition to its public health infrastructure, The Centre for Health Protection (CHP), which was created "to enhance existing disease control strategies and address new challenges."

Lessons learned in the SARS public health crisis focused on infrastructures that should have aggressively operated in the interest of various publics, but failed. That crisis required restated policies, new policies, new and refined administrative structures, and new infrastructures. A crisis quite often results when some organization (or, in this case, a complex of organizations) fails to exert the sorts of control that are expected by its constituencies. Once this failure by a central authority organization occurs, the ability of organizations linked in the system to exert their situational control also fails. This ripple effect emerges as the narrative of the crisis unfolds, and both lay and expert observers note how appropriate control measures were either not in place or were improperly implemented. As is the case of other crisis narratives, this one played out to the concern and anger of various stakeholding auditors who felt betrayed that the current systems failed.

The Health, Welfare, and Food Bureau indicated that "the Hospital Authority's eSARS system tracked all admissions at 14 acute-care hospitals, the Police's Major Incident Investigation and Disaster Support System (MIIDSS), an analytical criminal investigation tool, and the Department of Health's Case Contact Information System (SARS-CCIS)." Such detail is expected in crisis statements and is the ethical responsibility of those in decision-making positions. The narrative takes flesh once the characters demonstrate that they have good intention, the will, and means for reducing the recurrence of such crisis or related crises. Not only is an improved system needed, but the narrators of this story owe the auditors ample details to move the story theme from tragedy to happy ending. This ethical responsibility falls on all who are in positions of power. The document issued by the Health, Welfare, and Food Bureau announced other relevant details that could increase the perception that the organization knew and was able—as well as *willing*—to meet the requisite standards of control in the public interest.

Such changes and reports were especially necessary for crisis resolution by the key nodes in the system that had failed. This narrative, however, did not end there. Other systems reexamined themselves and worked for a higher level of crisis prevention and management. One such agency was

EUROPA—European Commission on Public Health. Its Web site described the symptoms of SARS and the primary and ancillary ways in which the virus can be contracted (sneezing and through exposure in laboratories). What has it done based on its efforts to refine itself?

> The Commission has taken a proactive and leading role in responding to this outbreak, co-ordinating [sic] surveillance through the EU Network or Communicable Diseases established under Decision 2119/EC/98, establishing an Expert Group on SARS and liaising with WHO, so as to ensure that Member States and Accession Countries respond in a coherent way to this outbreak. (EUROPA, 2004)

Such statements move the narrative to a higher plane to demonstrate care, concern, moral responsibility, and competence on the part of public health agencies. Mea culpas are one thing. Apology counts to the degree that it is accompanied by details that demonstrate that the key characters understand and move to enact increased control.

As we think of three stages of crisis and crisis response (precrisis, crisis, and postcrisis), we are reminded that, in fact, postcrisis responses are also part of precrisis communication. As one crisis works its way to successful conclusion, it reveals facts that are part of the information that lay and professional people want and need in the event that a similar crisis appears. Systems and protocols are so important in such situations that they cannot adequately and ethically be taken for granted. The collective management of risk requires that information relevant to the narrative of communicable disease be part of the story, at its outset, throughout, and in conclusion, which is actually the potential preparation for another segment in this drama. These narratives become the ongoing plot of public health intervention as communicable disease affects citizens' health and well-being.

By July 2004, WHO was able to report that the Ministry of Health of China had notified WHO of a SARS outbreak in Beijing and Anhui Province. WHO assisted local authorities to activate appropriate measures, including quarantine and rapid medical attention. The likely source of this outbreak of the disease was the National Institute of Virology in Beijing. Speedy response, including investigation, led authorities to recognize a pattern of spread among workers in one department. This research led to additional precautions in the way that biohazardous material is handled. One of these measures included following stringent international guidelines for validating the inactivation of live virus. Teams commissioned by WHO are now verifying compliance on this and other protocols. This narrative was quickly successful in bringing the epidemic to a halt.

Because of a failure on the part of public health authorities, two generations of contagion had occurred in 2002 instead of the success story of the intervention in 2004. Because of the cooperation and quick response by Chinese authorities in 2004, the epidemic and crisis of 2002 was not repeated. The protocols are correct.

CONCLUSION

The standard protocols of public health have evolved through practice and sound science. The key is to develop a narrative that meets the test of serving the risk bearers' interests. Optimal risk response assumes that a single, coherent, and functional narrative needs to be developed and shared by all parties in the most timely and expedient fashion. In the collective management of risk, the goal is maximizing the responses needed to mitigate damage as quickly as possible to reduce the adverse impact on the total population.

To this end, one narrative should emerge quickly that is based on sound science and offers professional and lay guidelines for mitigating risk. In the case of SARS, many competing and insufficient narratives arose. Thus, this inadequate, less-than-optimal response endangered the lives and interests of many risk bearers beyond what was needed and desired. As a narrative begins with "once upon a time" it should also speed to the resolution "and lived happily ever after."

REFERENCES

Barton, L. (1993). *Crisis in organizations: Managing and communicating in the heat of chaos.* Cincinnati, OH: College Divisions South-Western.

BBC News. (2004). Accessed September 14, 2004, from http://news.bbc.co.uk/1/ki/world/asia-pacific/2973415.stm.

Bechler, C. (2004). Reframing the organizational exigency: Taking a new approach to crisis research. In D. P. Millar & R. L. Heath (Eds.), *Responding to crisis: A rhetorical approach to crisis communication* (pp. 63–74). Mahwah, NJ: Erlbaum.

Bok, S. (1983). *Secrets: On the ethics of concealment and revelation.* New York: Vintage Books.

Bowen, S. A. (2002). Elite executives in issues management: The role of ethical paradigms in decision making. *Journal of Public Affairs, 2*(4), 270–283.

Bowen, S. A. (2004). Expansion of ethics as the tenth generic principle of public relations excellence: A Kantian theory and model for managing ethical issues. *Journal of Public Relations Research, 16*(1), 65–92.

Bowen, S. A. (2005). A practical model for ethical decision making in issues management and public relations. *Journal of Public Relations Research, 17.*

Burke, K. (1969a). *A grammar of motives.* Berkeley: University of California Press.

Burke, K. (1969b). *A rhetoric of motives.* Berkeley: University of California Press.

Campbell, A. (2004). The SARS Commission Interim Report: SARS and public health in Ontario. *Biosecurity and Bioterrorism, 2*(2), 118–126.

CBC News: Disclosure. Accessed July 8, 2004, from www.cbc.ca/disclosure/archives.

Coombs, W. T. (1999). *Ongoing crisis communication: Planning, managing, and responding.* Thousand Oaks, CA: Sage.

Douglas, M. (1982). *In active voice.* Boston, MA: Routledge & Kegan Paul.

Douglas, M. (1992). *Risk and blame: Essays in cultural theories.* London: Routledge.

DuHamel, C. (2003, August-September). Dealing with SARS: Toronto hospital learned to survive beyond the crisis communication plan. *Communication World, 20*(5), 26–30, 48.

EUROPA—European Commission on Public Health. (2004). Accessed July 8, 2004.

Feldman, S., Drache, D., Clifton, D., Fletcher, F., & Laing, A. (2004). *Media coverage of the 2003 Toronto SARS outbreak: A report on the role of the press in a public crisis.* Robarts Centre for Canadian Studies, York University

Fisher, W. R. (1984). Narration as a human communication paradigm: The case of public moral argument. *Communication Monographs, 51*, 1–22.

Fisher, W. R. (1985). The narrative paradigm: An elaboration. *Communication Monographs, 52*, 347–367.

Fisher, W. R. (1987). *Human communication as narration: Toward a philosophy of reason, value and action.* Columbia: University of South Carolina Press.

Fritsch, P., Pottinger, M., & Chang, L. (2003, April 7). Divergent Asian responses show difficulties in dealing with SARS. *Wall Street Journal-Eastern Edition, 241*(67), A1.

Hall, L. M. et al. (2003). Media portrayal of nurses' perspectives and concerns in the SARS crisis in Toronto. *Journal of Nursing Scholarship, 35*(3), 211–216.

Hearne, S. A., Hamburg, M. A., & Segal, L. (2004). SARS and its implications for U.S. public health policy: "We've been lucky." *Biosecurity and Bioterrorism, 2*(2), 127–131.

Heath, R. L. (1994). *Management of corporate communication: From interpersonal contacts to external affairs.* Hillsdale, NJ: Erlbaum.

Heath, R. L. (2004). Telling a story: A narrative approach to communication during a crisis. In D. P. Millar & R. L Heath (Eds.), *Responding to crisis: A rhetorical approach to crisis communication* (pp. 167–187). Mahwah, NJ: Erlbaum.

Heath, R. L., & Millar, D. P. (1994). A rhetorical approach to crisis communication: Management, communication processes, and strategic responses. In D. P. Millar & R. L. Heath (Eds.), *Responding to crisis: A rhetorical approach to crisis communication* (pp. 1–17). Mahwah, NJ: Erlbaum.

Holder, T. L. (2004). Constructing response during uncertainty: Organizing for crisis. In D. P. Millar & R. L. Heath (Eds.), *Responding to crisis: A rhetorical approach to crisis communication* (pp. 51–61). Mahwah, NJ: Erlbaum.

Hong Kong Department of Health. (2004). Press release issued May 25, 2004. Site accessed July 8, 2004.

Hong Kong health chief apologizes over SARS crisis handling. (2004, March 29). *Health & Medicine Week*, p. 750.

Johannesen, R. L. (1996). *Ethics in human communication* (4th ed.). Prospect Heights, IL: Waveland.

Kant, I. (1994). *The metaphysical principles of virtue* (J. W. Ellington, Trans.). Indianapolis: Cambridge. (Original work published 1797)

La Rocque, S. (2004). Site of United States Embassy, Tokyo, Japan. Accessed September 11, 2004, from http://japan.usembassy.gov/e/p/tp-20030520a5.html

Marra, R. J. (2004). Excellent crisis communication: Beyond crisis plans. In D. P. Millar & R. L. Heath (Eds.), *Responding to crisis: A rhetorical approach to crisis communication* (pp. 311–325). Mahwah, NJ: Erlbaum.

Maunder, R., Hunter, J., Vincent, L., Bennett, J., Peladeau, N., Leszcz, M., Verhaeghe, L. M., Steinberg, R., & Mazzulli, T. (2003). The immediate psychological and occupational impact of the 2003 SARS outbreak in a teaching hospital. *Canadian Medical Association Journal, 168*(10), 1245–1251.

McKie, D. (2001). Updating public relations: "New science," research paradigms, and uneven developments. In R. L. Heath (Ed.), *Handbook of public relations* (pp. 75–91). Thousand Oaks, CA: Sage.

Meng, X. (2003). *SARS crisis under globalization.* Retrieved May 6, 2004, from http://www.rieti.go.jp/jp/events/04011601/pdf/xiao.pdf

Nicholas, A. J. (2003). Using third-party advocates in a crisis. *Strategic Communication Management, 7*(5), 6–7.

Palmlund, I. (1992). Social drama and risk evaluation. In S. Krimsky & D. Golding (Eds.), *Social theories of risk* (pp. 197–212). Westport, CT: Praeger.

Pauchant, T.C., & Mitroff, I. I. (1992). *Transforming the crisis-prone organization: Preventing individual, organizational, and environmental tragedies.* San Francisco: Jossey-Bass.

Piovesan, R. (2003). Toronto's PR bungle on SARS. *Marketing Magazine, 108*(18), 7.

Seeger, M. W., Sellnow, T. L., & Ulmer, R. R. (1998). Communication, organization and crisis. In M. E. Roloff (Ed.), *Communication Yearbook* (Vol. 21, pp. 231–275). Thousand Oaks, CA: Sage.

Seeger, M. W., Sellnow, T. L., & Ulmer, R. R. (2003). *Communication and organizational crisis.* Westport CT: Praeger.

Sellnow, T. L., & Ulmer, R. R. (2004). Ambiguity as an inherent factor in organizational crisis communication. In D. P. Millar & R. L. Heath (Eds.), *Responding to crisis: A rhetorical approach to crisis communication* (pp. 251–262). Mahwah, NJ: Erlbaum.

Sriramesh, K., Grunig, J. E., & Buffington, J. (1992). Corporate culture and public relations. In J. E. Grunig (Ed.), *Excellence in public relations and communication management* (pp. 577–595). Hillsdale, NJ: Erlbaum.

Stacks, D. W. (2004). A multidimensional model of public relations. In D. P. Millar & R. L. Heath (Eds.), *Responding to crisis: A rhetorical approach to crisis communication* (pp. 37–49). Mahwah, NJ: Erlbaum.

Taiwan Security Research. (2004). Accessed September 11, 2004, from http://taiwansecurity.org/TSR-SARS.htm

Turner, B. (1976). The organizational and interorganizational development of disasters. *Administrative Science Quarterly, 21*, 378–397.

Weick, K. E. (1988). Enacted sense-making in crisis situations. *Journal of Management Studies, 25*, 306–317.

Yeoh, S. C., Lee, E., Lee, B. W., & Goh, D. L. (2003, June 21). Severe acute respiratory syndrome: Private hospital in Singapore took effective control measures. *British Medical Journal, 326*, 1394.

Zhao, Z. et al. (2003). Description and clinical treatment of an early outbreak of severe acute respiratory syndrome (SARS) in Guangzhou, PR China. *Journal of Medical Microbiology, 52*, 715–720.

9

Encephalitis Outbreaks and Public Communication Exigencies

A Metaphoric Analysis

Keith Michael Hearit

In mid-October 2003, public health officials in the Eastern United States were alarmed when the New York health commissioner reported that five individuals from Staten Island had been hospitalized and the culprit was believed to be the dreaded virus encephalitis. The victims, ages 22 to 54, presented with symptoms such as fever, headaches, tiredness, and a feeling of malaise; some even had seizures. Four of the patients were reported to be in critical condition (Singleton, 2003). As a result, a medical alert was issued almost immediately to physicians and others working in the public health community, urging them to be on the lookout for symptoms that might be found to be encephalitis. Public health officials and physicians marshaled their resources and rallied around the dreaded outbreak (Singleton, 2003).

The next day the alarm subsided, when the Department of Health and Mental Hygiene (DHMH) reported that the suspicion and fear of an encephalitis outbreak turned out to be unfounded and that the patient's illness was not due to an encephalitis outbreak ("Health Officials," 2003). Yet for many, the incident revealed critical concerns about the public health infrastructure and the ability of public health officials to identify and respond quickly to a public health crisis brought on by the encephalitis virus.

In effect, the alarm that sounded for public health officials demonstrated the necessity of effective crisis communication in a public health emergency. Historically, the study of crisis communication has focused primarily on organizations and their attempts to communicate to key publics and communities what they are doing to manage a problem or issue, whether the crisis involves being caught in a wrongdoing, as was Exxon, or being the victim of a psychopath, like Tylenol (Hearit, 2006). Increasingly, scholars have attempted to move the field of crisis management away from its corporate emphasis and relate it instead to other forms of organized external communication, whether by institutions seeking to come to grips with historical grievances, public school systems dealing with post-Columbine emergency planning, or governments and public health communities responding to crises that concern the health and well-being of communities, regions, and nations (see e.g., Millar & Beck, 2004; Reynolds, 2003; Reynolds & Seeger, 2005; Seeger, Sellnow, & Ulmer, 2003).

Consequently, the purpose of this chapter is to apply crisis management theory to the problem of public health crises, particularly as it relates to discourse surrounding the encephalitis virus, and offer comments about how the public health community communicates about the virus. As such, this chapter argues that the encephalitis crisis management discourse, and for that matter most exotic viruses and diseases, emphasizes the reduction of uncertainty. In particular, through a focus on the metaphors employed in the rhetorical construction of encephalitis, I show that public health communities use metaphors progressively, first treating the pending crisis as a battle that must be waged, and then, as the uncertainty is reduced and the risks from the virus are familiarized, concluding by treating it as a serious but manageable public issue. To explicate this thesis, this chapter (a) introduces the topic of encephalitis, (b) explicates the role of metaphor in the management of crises, (c) explains the method used to conduct this study, (d) explicates the metaphors used by public health officials in the management of encephalitis, and (e) concludes with a number of implications for crisis management in public health exigencies.

THE LOOMING THREAT OF ENCEPHALITIS

At its root, encephalitis is a virus and a lethal one at that. The term *encephalitis* actually is used to encapsulate a number of different forms of a single virus. The primary form of which most Americans are aware, and a significant focus of this chapter, is the West Nile virus. For most people, the West Nile virus is harmless, but in a few cases it can cause encephalitis, which is an inflammation of the brain that often is deadly and, if not deadly, may cause significant and permanent neurological damage. It is called the West Nile virus because it was first identified in the West Nile district of Uganda in 1937. It made its first appearance in the United States in 1999 (King, 2004; for more on the development of encephalitis and other emerging infectious diseases, see http://www.cdc.gov/ncidod/EID/index.htm).

Other forms of the virus are well known and equally deadly. Japanese encephalitis particularly affects Asia and the Far East; an outbreak believed to be Japanese encephalitis took the lives of 161 people in July 2003 in India ("Over 160 Die," 2003). Also lethal is encephalitis B; an outbreak in China in June 2003 had 296 reported cases, with 25 deaths ("Chinese Health Ministry Releases," 2003). Another familiar form of the virus is Eastern equine encephalitis, which is probably the most lethal form of the virus and is particularly deadly to horses. Other variations include St. Louis encephalitis, LaCrosse/California encephalitis, Meningo encephalitis, and Listeria-induced encephalitis. The virus is also related to the Herpes Simplex virus (HSV).

Encephalitis is an arbovirus, transmitted primarily through arthropods such as ticks or bloodsucking insects such as mosquitoes. It is a particular concern in the dryness of late summer, especially after a wet spring. The fact that it is not spread person to person is one major factor that tends to keep outbreaks small (Johnson, 2000; Purdy, 2000; Station, 2004; Weiner, 2004).

Encephalitis is indiscriminate; it attacks the young and old. People who are infected with the virus typically present with symptoms such as headache, fever, convulsions, altered consciousness, and signs of disorientation. Because encephalitis is a virus, there are no specific antiviral treatments available (antibiotics treat bacterial, not viral infections). As a result, health care professionals tend to provide ongoing nursing care and to treat the symptoms, such as the fever, with IV fluids, and to help with breathing. There often is an interval of several weeks between the advent of symptoms and the diagnosis.

Research into developing vaccines is underway, although given the relatively small numbers that characterize most outbreaks, there is not a huge financial incentive to do so. Two major vaccines currently exist—one for use against Eastern equine encephalitis (although it is for horses and not for humans) and the other for Japanese encephalitis (because it is especially prone to affect people in China).

Given that there is no real treatment for people who have the disease, public health efforts have focused on awareness and prevention, emphasizing the steps that individuals, counties, regions, and governments can take to limit the likelihood of an encephalitis outbreak (the precautions tend to be the same across all forms of the virus). Individuals, for instance, are counseled by public health authorities to wear long-sleeved shirts and slacks, use a mosquito repellent such as DEET™, and avoid being outside at dusk (30 minutes after the official sunset) when mosquitoes are especially active. Other frequently communicated advice is to drain any standing pools of water, restrict travel to areas where known outbreaks are likely to occur, and use care and caution in choosing recreational activities (such as hiking, where one might be especially prone to come into contact with ticks and mosquitoes). The other primary technique used by communities to prevent the spread of encephalitis, although somewhat more controversial, is spraying to wipe out mosquitoes in populated areas. These actions are consistent with larger principles of self-efficacy, which is a key aspect of many health messages (Egbert & Parrott, 2001; Witte, Meyer, & Martel, 2000).

In effect, because there is no real cure, the problem of encephalitis presents a difficult rhetorical situation (Bitzer, 1968; Farrell & Goodnight, 1981), in that, beyond prevention messages, communication about the disease is limited. The next section explicates the problem of uncertainty in crisis management and the role of metaphor in the reduction of that uncertainty. After addressing a number of methodological concerns, an explication is presented of the root metaphors in public health crisis communications that surround the encephalitis virus.

THE ROLE OF METAPHOR IN CRISIS MANAGEMENT

Due to its infrequent and unpredictable nature, perhaps nothing more directly characterizes the nature of a crisis than its uncertainty; indeed, uncertainty is featured in most crisis scholar's definitions. Coombs (1999), for instance, defined a *crisis* as "an event that is an unpredictable, major threat that can have a negative effect on the organization, industry, or stakeholders if handled improperly" (p. 2). Similarly, Crable and Vibbert (1986) described a *crisis* as "a rare kind of situation" (p. 254) created by both internal and external agents. Likewise, Fink (1986) argued that crises are "an unstable time or state of affairs in which a decisive change is impending" (p. 15). Seeger et al. (1998), in an extensive review of crisis management, concluded that a crisis is "a specific, unexpected and non-routine organizationally based event or series of events which creates high levels of uncertainty and threat or perceived threat to an organization's high priority goals" (p. 223).

Regardless of whether it is intentional, in effect all such views of crises, with their emphases on unpredictability and uncertainty, are rooted in a Weickian approach to conceptualizing crisis management. Writing on the topic of crisis, Weick (1988) argued that crises are "characterized by low probability/high consequence events that threaten the most fundamental goals of the organization" (p. 305). Against such threats, individuals and organizations organize through a process of enactment to make sense of and reduce the uncertainty that characterizes their environment (Weick, 1979, 1988). This sensemaking is a process that precedes the interpretation of uncertain situations. In other words, sensemaking is a reflective process whereby human beings force stimuli into meaning frameworks—that then render such information meaningful (Seeger et al., 2003; Theus, 2004).

Crises are, by their nature, equivocal situations. Social actors "attain at least a partial consensus on the meaning of their behavior and that they look for patterns that underlie appearances, actions, and events" (Weick, 1988, p. 165). One way to look for these underlying patterns is through the use of metaphor, which are reality-creating vehicles or sensemaking devises by which crisis managers come to understand and make sense of the uncertainty.

Metaphor Defined and Applied

At root, the characteristic of a metaphor is that it brings about an understanding of an idea, topic, or object through comparison. Lakoff and Johnson (1980), who published the seminal *Metaphors We Live By*, explained: "The essence of metaphor is understanding and experiencing one kind of thing in terms of another" (p. 5). Indeed, this notion of using the familiar to explain the unfamiliar is present in most definitions of metaphor. Kendall and Kendall (1993) defined a *metaphor* as "saying that one object is another" (p. 150), whereas Brown (1987) defined *metaphor* as "abstract qualities perceived as similar from dissimilar phenomena" (p. 12). Said another way, metaphors are compact images embedded in experiential language; yet without context, metaphors have no meaning.

Although people often conceive of metaphor as simply stylized or decorative language, the primary way that people think is metaphorical in nature (Propp, 2003). Kendall and Kendall (1993), for instance, argued that metaphors are "intimately interconnected with the way we think" (p. 149). In structuring how people think, metaphors actually help people socially construct reality (Hearit & Courtright, 2003; Orr, 1978). As such, it must be noted that not only do they structure how people participate in reality, but they may also help people to create new realities. Lakoff and Johnson (1980) summarized it in the following way: "The way we think, what we experience, and what we do every day is very much a matter of metaphor" (p. 3). In other words, people use metaphor, in a sensemaking way, to organize and classify, as well as make sense of uncommon situations.

Metaphors and Crisis Management

Yet if metaphors provide a way to name and frame an uncommon situation, it must be recognized that such a name privileges one point of view or course of action over another. As Burke (1966) acknowledged: "Even if a given terminology is a reflection of reality, by its very nature as a terminology it must be a selection of reality, and to this extent it must function also as a deflection of reality" (p. 45). The same holds true for the use of metaphor as it is applied to crisis management. A metaphor can function to reduce uncertainty by telling crisis managers how to act in a particular situation. In so doing, the uncertainty reduction occurs in that metaphor simplifies a complex and multifaceted crisis context into an easily understandable state of affairs.

The use of a metaphor is also evaluative, not merely decorative (Hayakawa, 1978). It "is a driver affecting both the choice of what decisions to make and how those decisions are made" (Lissack, 1997, p. 295). Metaphors create for organizations and institutions the range of ideas on how to potentially act, as well as how they ultimately will act. As Propp (2003) argued, "Metaphors are fundamental in shaping our reality and the behavior that ensues from those perceptions" (p. 113). Subsequently, crisis managers' metaphor use legitimizes their actions (Millar & Beck, 2004).

In a metaphoric analysis of crisis discourse, Millar and Beck (2004) concluded that there are three principal crisis metaphors that account for the bulk of the organization's language use during crisis management: war metaphor, container metaphor, and disease metaphor.

These metaphors do not simply represent one term, but often are characterized by a whole cluster of terms (Ivie, 1987). Although some are directly evident ("We're at war against this"), others are more nested ("How did we get into this mess?").

War Metaphor. Probably the prevailing metaphor of all crisis management discourse is that of war. Such an approach characterizes crisis management as a battle to be won or lost. In this turbulent and often hostile environment, the times are no longer normal, but are critical and stressful; because of this fact, organizations and institutions must marshal their existing resources against a dangerous enemy (with negative characteristics) in the pursuit of a defined objective. To be victorious, wars require leaders; as a result, use of this metaphor envisions organizations constricting and formalizing their decision making and communication (Billings et al., 1980).

Wars are exhausting, however, and concern must be taken to ensure that the negative implications of the metaphor do not come to fruition. There is a danger that use of this metaphor may create a siege mentality, disabling the organization and preventing the return to a normal state when the crisis is over—or, at minimum, prohibiting the organization from seeing other potentially positive courses of action.

Container Metaphor. Although a war metaphor is easy to identify, a container metaphor is a bit more obscure, but nevertheless quite useful in conceptualizing a crisis. A state or an activity is viewed as a container (i.e., how did you "get into" that context?). In this way, a container metaphor draws clear boundaries of that which is inside from that which is outside. In a crisis situation, outside forces have broken the container (i.e., an iceberg or reef breaks through an oil tanker) or an inside agent causes a serious problem—which allows events to get out of control and induces fear. As such, the container metaphor suggests that some unequivocal situation has become equivocal due to the penetration of the inside from the outside. A break in the container suggests that no one is "in charge" (Millar & Beck, 2004, p. 162) and, as a result, accounts for much of the uncertainty that characterizes crisis management situations. Hence, conceptualizing a crisis from this metaphorical perspective means a loss of control, a break from the normal routine, and crossing a turning point from order to disorder. Millar and Beck (2004) noted that a war metaphor is also a container metaphor.

Disease Metaphor. The third type of metaphor is that of a disease. Fink's (1986) early model of crisis is rooted in such an idea, particularly his use of the term *prodrome*, a medical term for a symptom of a disease. This metaphor harkens back to the original conceptualization of the term *crisis*, which Seeger, Sellnow, and Ulmer (2003) note comes from the Greek *krisis*, which "was used as a medical term by Hippocrates to describe the turning point in a disease, and from *krinein*, meaning to judge or decide" (p. 7). A crisis conceptualized as a disease metaphor, then, is one in which the organization should be on the lookout for the symptoms that connote the onset of the disease and must marshal its antibodies to fight off the external agents that have invaded it to return to its preferred homeostatic state. It should be noted that the language used for a disease metaphor is virtually indistinguishable from that of a war metaphor.

METHOD

As Embler (1966) suggests, "It is perhaps on the linguistic edges, on the intellectual fringe, where we find the kind of language that tells us most about ourselves as a people" and that can "reveal hidden wishes and ideals" (p. 31). The analysis of the metaphors used by crisis managers and journalists when conceptualizing a crisis provides a window (itself a metaphor) into their sensemaking, as well as offers a vehicle by which to understand their actions and behaviors. This is particularly useful in that the meaning of the metaphor is rich particularly because those who use metaphor are likely not "linguistically conscious" of it (Embler, 1966).

To complete this analysis, a metaphoric analysis was performed using Lexis-Nexis from January to February 2005. In particular, to understand the primary metaphors evident in *The New York Times* encephalitis reporting, the following search terms were entered: encephalitis, West Nile virus, and the accompanying search terms: battle, campaign, crisis management, exotic, first cases, frontlines, infectious disease, management, outbreak, rare, soldiers, surveillance, and tracking. *The New York Times* was chosen because it is the closest thing we have to a national newspaper and because it has been shown to greatly influence newspaper coverage throughout the nation (Batulis, 1976; Beniger & Westney, 1981; Charles, Shore, & Todd, 1979). However, because not enough reports were made to warrant a study, the search sources were broadened in Lexis-Nexis to include national and international news sources. The use of the search terms revealed the presence of 54 usable articles by which to assess the use of metaphor.

This study draws from Koch and Deetz (1981) in laying out the methodology for assessing the use of metaphor. In the first step, all of the stories were read and the metaphors present in the reporters' stories identified. Second, these articles were systematically assessed and analyzed for specific metaphors. Next, these metaphors were arranged into clusters. Three metaphoric clusters were found to be present: an exotic metaphor, which sees the virus as an alien enemy; a surveillance metaphor, which views the virus as a serious threat to be tracked; and a management metaphor, by which the virus is seen as a manageable problem.

ANALYSIS OF CRISIS METAPHORS

As would be expected from the review of the crisis metaphor literature, the use of a military metaphor predominates in the encephalitis public health crisis reporting. It certainly is the central way in which people name, classify, and describe the threat of encephalitis. However, on close consideration, how the metaphor is enacted undergoes remarkable evolution.

Exotic

The first set of metaphors reveals the presence of encephalitis using an exotic/distance metaphor. In particular, this set of the stories views encephalitis as "over there" and rare, but certainly on the march here. This is evidenced by the fact that most of the reported outbreaks are overseas. Given that the virus is unfamiliar, the emphasis is on how little is unknown and how unsure the public health community is about the virus; accompanying this uncertainty is a profound sense of fear and dread over what is coming. One way to reduce the uncertainty is the invocation of the influenza flu pandemic of 1918–1919 as an example (Barry, 2004). By relating the virus to a known earlier pandemic, scholars, reporters, and public health officials are able to reference and predict what would happen in a contemporary outbreak of an infectious disease.

Part of what makes encephalitis exotic is where it originates. By and large, most encephalitis viruses originate in Asia, especially in the Guangdong province of China. In 2003, for instance, the Associated Press reported that "an encephalitis outbreak in southern China has killed at least 24 people and infected more than 300" ("Encephalitis Outbreak Kills," 2003, n.p.). The use of the term *outbreak* is a military term which suggests a dangerous and sudden occurrence of fighting. Yet Guangdong health officials characterized such numbers as normal for China, in that encephalitis B "kills nearly 10,000 people each year in Asia, most of them children, and it infects 40,000" ("Encephalitis Outbreak Kills," 2003, n.p.). The fact that such numbers of deaths and infections is unimaginable in the West also adds to its foreign nature.

Other stories about the outbreak, in addition to the emphasis on casualties, also referenced military terminologies in describing how Asian health authorities were seeking to fight the virus. One report of the outbreak used a martial metaphor, describing it as an enemy on the march, noting that it "starts to appear in south China around June, gradually moving north over the following month" ("85 More Infected," 2003). Another report about efforts in Hong Kong to prevent the virus from reaching the megalopolis discussed how the virus had "flared up" and spread to "urban territory," and that, as a result, Hong Kong health officials had "tightened precautions" and "stepped up surveillance," as well as emphasized the need for "early alerts" ("Viral Encephalitis," 2003, p. 20). Given the presence of the martial efforts, many reports detailed the steps that public health officials were taking to fight the virus. Because there is a vaccine for encephalitis B, much of the efforts to counteract the virus usually take the form of making sure that those most at risk are vaccinated. As such, reports suggested that efforts to administer the vaccine as well as that of "eliminating mosquitoes" "are in full swing" (Chinese Health Ministry, 2003, n.p.). Given that fighting an enemy is not just about technology, but about character, another report of the outbreak was met with a "renewed pledge" to "fight the disease that has killed 24 people in Guangdong" ("77 People Struck Down," 2003, n.p.). The efforts were also accompanied by the formation of a "special task force," as well the development of a strategy to "wipe out mosquitoes" ("77 People Struck Down," 2003, n.p.).

Adding to the sense on the part of many that encephalitis constitutes an alien threat was the fact that 2003 was also the year of a severe outbreak of encephalitis in India. This outbreak "claimed at least 161 lives" in the states of Andhra Pradesh and Assam ("Over 160 Die," 2003, n.p.). Although experts were unsure which form had caused the deaths, some believing it to be Japanese encephalitis, whatever its form, news accounts reported that the virus had been on a "grim march" and was "taking the toll" on the lives of

Indians ("Over 160 Die," 2003, n.p.). The fact that the Indian health officials were unable to definitively determine which encephalitis virus had caused the outbreaks adds to its exotic nature, as well as frames the virus as an enemy that is so dangerous it can even change its form.

Due to the fact that the knowledge and understanding of encephalitis and many other infectious diseases is so limited, some of the reports draw direct comparisons of contemporary problems with the influenza flu pandemic of 1918–1919 (Barry, 2004). This allows people to understand an unknown threat in terms of a previous known virus—although even this metaphorical application is not reassuring. These reports reference how widespread and deadly a serious outbreak could be—even in these medically advanced times, given that it is not cost-effective for vaccines to be developed for every known virus. Barry (2004), for instance, reported that in 1999, "the Centers for Disease Control and Prevention conducted a study of what would happen in the United States if a new influenza pandemic struck. It concluded that the death toll most likely would range from 89,000 to 300,000, but there was a one-in–20 chance that the deaths could exceed 400,000" (p. 13A). What made the study so chilling was the fact that it assumed that a vaccine was available and that the major problem facing public health officials was one of production and distribution of the vaccine.

Another facet that makes encephalitis exotic is that it constitutes a new threat for the West. For instance, many have watched the West Nile virus in its march across the country, starting in New York in 1999 and making its way to the Western United States in 2003. In discussing the newness of the threat, Simhan (2004) wrote that Californians were living in fear of West Nile. "This is partly because earthquakes, unlike West Nile virus, aren't a new risk" (Simhan, 2004, p. E1). Specifically, he noted that the "hubbub about West Nile has dropped back East—where cases first appeared in 1999—but anxiety is fresh here because California did not have any cases until last year," noting that the chances of dying from West Nile has risen from "1 in 520,000 in 1999 to 1 in 15,000 in 2003" (Simhan, 2004, p. E1). In this way, the referencing of the odds of getting the virus are often presented with a lottery metaphor.

Finally, part of what makes encephalitis exotic is that, although the public health community is making preparations to deal with the virus, it is not yet ready. One news account, for instance, described the West Nile threat as one that people "don't understand or have control over" (Johnson, 2000, p. 27), and that although they fear the virus, they also fear the widespread use of pesticides to fight it. Another article reported that Dr. Julie L. Gerberding, the director of the Centers for Disease Control and Prevention (CDC), told the Fourth International Conference on Emerging Infectious Diseases that most governments are not prepared to deal with outbreaks like encephalitis and other infectious diseases (Altman, 2004). Dr. Keiji Fukuda, a CDC specialist in epidemiology, reinforced this view when he admitted,

"We are thin in terms of numbers of available experts, resources and staff" (Altman, 2004, p. F6). Specifically, he observed that the recent outbreaks of infectious diseases "have taxed and strained the entire international health system" (Altman, 2004, p. F6).

Early efforts to counteract the encephalitis virus view it as a battle to be joined against a deadly and exotic enemy that systematically works its way through a region in a manner much like a military campaign, and that technology (vaccines) must be marshaled to counteract the disease. It is important to note that such a view does not permanently predominate. Instead, as the virus reaches people's doorsteps and they come to deal with it on a more frequent basis, the metaphor changes from an exotic war metaphor to a surveillance metaphor. Hence, the accepted wisdom that surrounds the use of a surveillance metaphor is that encephalitis is still dangerous. However, given its arrival where people live, it comes to be seen as a familiar threat that must be closely watched.

Surveillance

The second set of metaphors put emphasis on a surveillance theme, emphasizing watchfulness and vigilance. This occurs when the first cases of encephalitis are reported in the United States. In particular, stories tend to focus on the first in the county, region, or state to deal with the problem of encephalitis. As the virus arrives, nervous public health officials report that there is still no known cure. Because there is no cure, the emphasis is particularly on tracking and following the virus. As a result, people are encouraged to take precautions to avoid contact with the virus. This also helps to reduce the uncertainty that revolves around the virus.

The use of the surveillance metaphor often means that what was an "over there" kind of problem now requires surveillance because it is on its way here or is here. Most people's experience with encephalitis was when the West Nile virus arrived in 1999. Reports viewed it as a new threat, but one that was likely to follow a somewhat familiar script. One *New York Times* article described the region as "in the grip of another threat that combines the specter of disaster with what may eventually be seen as overwrought drama" (Purdy, 2000, p. 21). As such, the need for surveillance could not be over-emphasized. A Rockland County, New York, official, Cott Vanderhoef, for instance, described the threat of West Nile as "a moving target in the night" due to the fact that "There's been no one who's been able to tell this county executive how big a threat this is to humans" (Purdy, 2000, p. 21). The purpose of the surveillance was for mosquito control. The story noted that spraying had begun, but that the emphasis was on killing mosquito larvae throughout the year, rather than just spraying to kill adult mosquitoes. Similarly, Purdy noted that "Bergen County has increased its surveillance for West Nile virus" (Purdy, 2000, p. 21). The reason for doing the spraying, of course, was protection of the citizenry; Thomas Herzog, a township supervisor in Westchester County, said, "I think the powers that be don't want to have the accusations come back that they were negligent in protecting the public health" (Purdy, 2000, p. 21).

Tracking the disease continued to be an ongoing effort. A report from 2003, for instance, emphasized the need for better tracking of the disease. The Associated Press reported that the Bill & Melinda Gates Foundation had awarded a $27 million grant to fight Japanese encephalitis in Southeast Asia ("Gates Gives $27M Grant," 2003). The purpose of the grant was to fund "disease surveillance" until a vaccine can be developed (n.p.). In commenting on the grant, Thomas Chambers, a molecular microbiology and immunology professor at Saint Louis University, argued, "I think this is the time people are considering seriously controlling JE as a virus" ("Gates Gives $27M Grant," 2003).

Of course, one of the primary benefits of increased surveillance is the fact that people become much more familiar with the virus and, as a result, uncertainty is greatly reduced and people are less fearful. After the virus had been in the United States for 3 years, for instance, there was a sense that, although the virus was a major concern, it was not likely to become a crisis. In one news report, Dr. Julie Gerberding, head of the CDC, said that West Nile "has been an evolving problem, so therefore there is a certain sense of urgency in making sure the public health response is optimized. But I don't think we need to send the message there's a crisis. There is no need to panic" (Altman, 2002, p. A13). She reminded people that the best way to control for West Nile is to spray to control mosquitoes. Yet although spraying mosquitoes presents a way to fight the disease, it is not foolproof; ongoing surveillance is necessary. Gerberding "cautioned that quelling an outbreak of mosquito-borne disease is complex and includes unknowns like the weather. The interaction of the size of the mosquito population, geography and the weather could make the problem worse or better" (Altman, 2002, p. A13). James Hughes, an infectious disease specialist with the CDC, reinforced the need for surveillance, warning that all mosquito-borne diseases are a problem: "The key to preventing their return as a public health threat in this country was to stay vigilant" (p. A13). As to where West Nile will be a major problem, Hughes said: "It will take five years to get a clearer answer" (p. A13). Surveillance promotes planning and being prepared. It also is part of the established methodology of epidemiology as disease detective.

Indeed, with the ongoing reporting about the virus, people become much more acquainted with it in all its forms, again reducing the uncertainty that revolves around it. West Nile, for instance, will no longer be novel, but will become part of the routine set of seasonal threats. A September 2003 report, for instance, detailed the fact that Louisiana was now facing three different types of encephalitis threats (McConnaughey, 2003). The bulk of the threats were from West Nile, but there were also cases of St. Louis encephalitis and Eastern equine encephalitis. Mississippi also reported significant problems with the disease. Part of the issue was how closely states track the disease. Dave Daigle, CDC spokesperson, said, "There may be more in some states, or it may be just that some states are very aggressive in tracking it" (McConnaughey, 2003, n.p.).

Yet the fact that the virus is becoming more familiar as it is being tracked does not make it any less dangerous; it is still a serious threat. The dangerous nature of the disease is still emphasized in the surveillance stage, as is the fact that there is no real treatment for encephalitis. For example, there were outbreaks in both New York and New Jersey in 2003. In Suffolk County, New York, three Montauk area parks were closed due to the threat of Eastern equine encephalitis after mosquitoes that carried the virus were sampled there. Eastern equine encephalitis is more dangerous than West Nile and kills "about 35 percent in the people who contract it; about half of those who survive have neurological damage" (Werde, 2003, p. B5). Likewise, New Jersey health officials reported the first cases of Eastern equine encephalitis in almost 20 years—in which a 2-year-old girl was "stricken with a fever and seizures," but was likely to recover (Kocieniewski, 2003, p. B5). The virus killed six horses. Eastern equine encephalitis also caused a larger outbreak the following year (2004) in New York State. Although no one died, the outbreak had "claimed up to 10 horses bitten by infected mosquitoes . . . a year after the virus took its largest human toll on record in the United States" (Weiner, 2004, p. A1). Given that the disease is so rare, officials did not seek to manage it with a vaccine or other methods. Instead, New York State officials argued that the best defense was to rely on their ongoing campaign to "Fight the Bite," which educated people on the dangers of all mosquito-borne diseases.

The use of a surveillance metaphor resulted in news reports that tracked the locations of particular outbreaks. A report of Eastern equine encephalitis, for instance, detailed a problem in Southeastern Massachusetts that showed that health officials expected the region to be "hit hard" by the virus (Lazar, 2004a, p. 2). In particular, the virus was found in Middleboro, an area that public health officials described as a "hot spot" (Lazar, 2004a, p. 2). Also, the appearance of the virus so early in August was seen as a particularly "bad omen" (Lazar, 2004a, p. 2). Similarly, in 2004, California officials noted problems with the virus due to its early appearance, which Hall (2004) described as a "bad omen" (p. A6). In a continued use of the martial metaphor, Hall (2004) noted that health experts "have been tracking West Nile's inexorable spread" (p. A6) since 1999. As a result, California officials had set up a "surveillance program" to watch for "increased activity" (Hall, 2004, p. A6).

The purpose of surveillance, of course, is not strictly informational; instead, surveillance has a practical purpose—typically for informing of the pattern of spraying for mosquitoes. In one article in the *Boston Herald* in 2004, the headline used a military term (as well as a colloquial one) that gives the outcome of surveillance: Swat teams target skeeters (Lazar, 2004b). Specifically, Lazar (2004b) wrote: "Worried that weekend rains will spawn an army of disease-carrying mosquitoes, health officials are declaring war on the bugs with stepped-up spraying and tracking to halt the spread of deadly eastern equine encephalitis . . ." (p. 5). The assault occurred after a 64-year-old man was infected by mosquitoes.

As understanding of the encephalitis virus becomes more complete, surveillance methods also become more sophisticated. One story, for instance, discussed a surveillance program in Washington state, which was one of only four states without a human case of the West Nile virus. One of the methods for detecting the disease is to track songbirds that can carry the virus. Although different birds react differently to infection, one *sentinel* of the virus is crows (King, 2004). Crows, according to King County officials, "die about five days after infection and have been important sentinels for the disease in other cities" (King, 2004, p. A1). A similar program in another part of the country emphasized the military-like surveillance in the battle against West Nile, albeit this time with chickens, not songbirds. One news account carried this headline: "A few good hens drafted in the war on West Nile," reporting a story about so-called sentinel chickens, which like crows are considered an early warning system for West Nile (Grissett, 2004, p. A1). One of the reasons chickens were chosen for this "noble service" (p. 1) is that they are especially resilient—the virus does not kill them, nor do they transmit it to others (unlike songbirds). "And if that's not enough to qualify these Rhode Island Red pullets for front line guard duty in the escalating war against mosquitoes, Heurtin [a mosquito control officer in Jefferson Parish, Louisiana] said Thursday that chickens are just plain tough birds" (Grissett, 2004, p. 1).

The emphasis on surveillance serves as an important step in reducing uncertainty. Yet it is only as reporting moves to a final metaphor do people come to see encephalitis as not a health care crisis that must be closely tracked, but rather an important problem that must be managed.

Management

When encephalitis finally arrives, there is the presence and emergence of a management metaphor—one that presents the virus as more familiar. In this use of metaphor, there is an emphasis on resignation, assimilation, and control. As a result, there is an awareness that the impact of the virus is not as bad as expected, but there still needs to be an accompanying emphasis on prevention strategies—all which fit with the management metaphor. There is also the emergence of a language for discussing the disease, and the metaphors that are used to explain it are much more familiar. This management metaphor also serves as the primary justification for county officials to seek to implement mosquito spraying over populated areas. Of course, it is no small stretch to view the mosquito control planes as fighter jets in the battle against the virus. In this way, disease management is also part of a larger medical metaphor.

One of the interesting features about this third stage is that there is the appearance of activists for the purpose of the management of the virus. For instance, one organization that emerged from concerns about the virus was Encephalitis Global, an organization founded by Wendy Station, a survivor of an encephalitis attack. The purpose of the organization is to serve as a clearinghouse of information about the virus, its management, and support for its victims. One of the frustrating facts, according to Station, is that, due to the military metaphor, media stories tend to focus on casualties, rather than survivors and the lives they live (Station, 2004).

Similarly, at the management stage, there is also the emergence of institutions to seek control of the virus. In particular, the institutional response is twofold. The first step in managing the virus is spraying to kill mosquitoes and mosquito larvae. This can be seen in a 2004 campaign in California in which many of the efforts centered on management of the virus, although still referencing a military campaign metaphor. Conway and Hoskins (2004) wrote, "As the West Nile virus marches across America, valley health officials are marshaling their forces for a pitched battle this summer" (p. B1). Yet while they drew from a martial metaphor, the emphasis was not on a fight, but on management. They then quoted Merced County California public health officials, who said, "The main thing we in public health want to emphasize is that we want widespread awareness. There's no need for panic, but the best thing anyone can do is prevent bites by mosquitoes to themselves and family members" (Conway & Hoskins, 2004, p. B1). The story then proceeded to explain the virus in great deal, as well as the steps people can take to prevent its transmission, further functioning to reduce uncertainty. Management, with its emphasis on self-efficacy, also denotes a sense of control.

The second facet of the institutional response is informational. There is the disclosure of a great deal of information on managing—and preventing contact with the virus—to further reduce uncertainty. A campaign metaphor is often present—a military metaphor in some sense, but much more precisely a public health/public relations campaign. For instance, Doa Ana County in New Mexico began a regional antimosquito campaign in May 2003. This informational campaign exhibited mosquito traps for use to catch mosquitoes for testing, the use of mosquito larvae-eating fish for use in stock tanks and water ponds in backyards, and an amnesty program for property owners who had old tires lying around (Hopkins, 2003).

This emphasis on a public relations campaign has a subsidiary effect: The bulk of the reporting tends to center around the personal efforts that individuals can take to avoid contact with the virus and, in particular, mosquitoes. One of the key ideas becomes one of vigilance (Hathaway, 2004). Jefferson County, Missouri, health officials, in revealing a case in which a 19-year-old woman had contracted a case of West Nile, emphasized that, in addition to the efforts at spraying by the county, there were many steps that individuals could do to prevent the spread of the virus. To avoid contact with mosquitoes, for example, people were encouraged to dress in long pants and shirts, wear mosquito repellent, and eliminate any potential standing areas of water such as bird baths, eaves troughs, and old tires. Steven Crawford, of the Jefferson County Health Department, reported that most people were willing to do their part in managing the virus: "Most people are more than willing to do something; they just need some guidance" (Hathaway, 2004, p. 1). Similarly, a California story tended to focus on the efforts people can take, noting that, "in the war against the biting bugs, these precautions can be effective weapons" (Griffith, 2004, p. A1). Recognizing the shift from worry to resignation, the article points out that the simple recommended steps include: "wear insect repellent, fix screen doors and windows, drain standing water and stay indoors when mosquitoes are busiest—they are often neglected" (p. A1).

These shifts in uncertainty even have an effect on the reporting. A *New York Times* article reported the change that has occurred from when West Nile first arrived as people have come to understand and adapt to the virus. Perez-Pena (2003) wrote: "In our metropolitan area, where the West Nile form of mosquito-borne encephalitis first made the leap from Old World to New in the summer of 1999, it has come to feel ho-hum—quite a change from the great attention and anxiety it drew back when" (p. 41). Furthermore, Perez-Pena reported on an analysis conducted by the paper, which showed that the number of West Nile stories by journalists fell by half and were less urgent, shorter, and not as prominently featured as in earlier years.

Perhaps the most interesting implication of the management metaphor is that it takes a risk that once was a major fear on the part of many and turns it into a bureaucratic funding issue. In other words, the evolution of encephalitis, with the heavy emphasis on a martial metaphor evolves into a problem that must be managed in an ongoing fashion. In Ed Dorado County, California, in 2004, for instance, there was a fight about whether or not to spend money on a survey to determine whether people were willing to support a tax to fund mosquito eradication efforts. This left public health officials incredulous. Joe Harn, a county official, exclaimed, "I cannot articulate my frustration over this matter. You don't ask for an appropriation to kill mosquitoes—you ask for an appropriation for a survey. This is Bureaucracy 101" (Locke, 2004, p. N5). The amount of the assessment was expected to be about $8. A similar problem occurred in Ontario, Canada, when the provincial government cut West Nile funding by 22% for the 2004 year (Mallan, 2004).

IMPLICATIONS OF A METAPHORIC ANALYSIS OF PUBLIC HEALTH CRISIS MANAGEMENT

This chapter has analyzed the discourse on the topic of encephalitis and how it is enacted in the management of the virus by public health officials. Specifically, it has argued that the encephalitis discourse is paradigmatic, in that it demonstrates how individuals and institutions in the health care community can use metaphor in dealing with uncertainty about looming threats. In so doing, it has found that war is the predominant metaphor used by public health officials—with the public health community operating on the frontlines doing battle with an insidious enemy who is not completely understood and is capable of quickly changing tactics. As a result, public health officials as well as journalists typically recommend caution as well as careful and strategic action against the enemy.

In examining the encephalitis discourse, a number of conclusions become clear. A primary outcome of this study is that it offers a support to Reynolds and Seeger's (2005) model of Crisis and Emergency Risk Communication (CERC). As with Reynolds and Seeger, this study has shown that encephalitis public health crises do "develop in largely predictable and systematic ways . . ." (p. 51), such as from risk and fear to outbreak and finally to resolution.

This study not only confirms Reynolds and Seeger (2005), it also suggests that metaphor is a primary influence on people's perception of the disease, as well as how they communicate throughout the process. This is supported in three ways. First, this study suggests that in the precrisis stage, the role of metaphor further explains how people conceptualize threat in the CERC model, in that people tend to articulate threats in foreign or exotic terms. Second, the emphasis on uncertainty reduction in the initial event stage of the CERC model in particular is highlighted, in that people use martial, surveillance, and ultimately management metaphors to make sense of encephalitis and, one would suspect, of most public health crises. Metaphors enable individuals and institutions to make sense of public health threats; that is, they translate them from foreign adversaries that they have no other reaction to than fear into a concept for which they have a familiar frame of reference. Finally, metaphor is present in the maintenance stage of the CERC model, in that a management metaphor is a familiar one by which people are able to arrive at a certain measure of comfort with the foreign agent as well as take steps of self-efficacy by which they can have a sense of control.

Although a martial metaphor clearly predominates in the discourse related to this problem, there is an evolution in the metaphors that characterize this form of discourse. In particular, the discourse tends to progress from one that, although still militaristic, tends first to treat the disease as a vilified foreign enemy. As the virus draws closer and individuals and institutions directly experience the virus, the metaphoric emphasis turns to that of threat surveillance and then management. In other words, as demonstrated by this analysis, typically people view encephalitis as an exotic enemy, but then come to realize that it is just another risk, like the many others in their lives (e.g., too much sun, too much greasy food). As with other risks, to control it, they must adapt and modify their behavior. In this way, the evolution tends to progress from a straight military metaphor to a container metaphor.

Accompanying this change in metaphor is one of tone and emphasis. At first, as demonstrated in how they use metaphor to discuss encephalitis, people view the virus as one that cannot be controlled and should be feared. In effect, an exotic orientation to encephalitis is scenic oriented in its worldview (Burke, 1969; Ling, 1972). But as people become comfortable with the encephalitis problem, there is a change to an agency-agent-centered ratio, which presents institutions and their public health machinery (agency) coming to grips with the threat it constitutes and formulating an organized response (typically in the form of mosquito control spraying), but also in terms of messages that demonstrate the need for individual response (agent), such as taking the necessary precautions to avoid the virus (e.g., getting rid of standing pools of water, wearing long-pants and shirts, and wearing DEET™).

Given the emphasis in this chapter on the use of metaphor, risk and crisis, managers should take great care and caution in how they describe an emerging infectious disease. In particular, they should be careful to avoid the temptation to describe a disease or virus in exotic and xenophobic terms—which causes people to fear others. Instead, issue managers should admit that, although viruses are different, none has absolutely no prior antecedent—and the vehicles and methods for identifying, tracking, and managing them epidemiologically and in a public health context have not changed. Instead, as the CERC model suggests, metaphors should be used that emphasize the self-efficacy health consumers can and should take when confronting viruses such as encephalitis (Egbert & Parrott, 2001; Witte, Meyer, & Martel, 2000).

In many ways, this analysis shows how encephalitis is a metaphor for how people fear that which is foreign and unknown—and that that fear is rooted in uncertainty. As such, this chapter has implications for a wide range of infectious diseases such as AIDS. People project a great deal of xenophobic alarm onto emerging infectious diseases and often provide them with anthropomorphic characteristics. Given the lack of familiarity on the part of most with regard to these diseases, many efforts reach for a metaphor that describes them as exotic and different from any other virus people have encountered. As a result, to describe them in terms that people can understand, there often are references to the 1918–1919 influenza pandemic, which killed millions. This happens despite that each emerging infectious disease is rooted in its own epidemiology, often with different vectors of transmission. Yet in each instance, as scientists, public health officials, reporters, and health care consumers come to understand a particular virus, the progression is likely to move from metaphor that emphasizes the exotic to one that favors the familiar.

Although the use of the war metaphor is in some ways valuable in attempting to get policy members to support and adequately fund public health initiatives to combat the disease, as well as quickly develop programs in which governments seek to track and manage encephalitis, one also has to wonder about the negative consequences of such a metaphor. For one, this war against infectious diseases is likely to be a war that never ends. Furthermore, one also suspects that this use of the war metaphor will one day be used in a public health content to justify restrictions on civil liberties—should and when a major outbreak ever occur.

ACKNOWLEDGMENTS

The author would like to thank Kathleen Propp for her gracious assistance in the development of this chapter.

REFERENCES

77 people struck down with encephalitis B. (2003, June 28). *China Daily*, n.p. Retrieved January 21, 2005, from Lexis-Nexis.

85 more infected with encephalitis, but China says outbreak under control. (2003, June 30). *Agence France Presse—English*, n.p. Retrieved January 21, 2005, from Lexis-Nexis.

Altman, L. K. (2002, August 6). West Nile outbreak is a warning, not a crisis, officials say. *The New York Times*, p. A13.

Altman, L. K. (2004, March 9). New microbes could become the"new norm." *The New York Times*, p. F6.

Barry, J. M. (2004, February 10). History offers lessons on flu's threats. *USA Today*, p. 13A.
Batulis, A. P. (1976). Congress, the president, and the press. *Journalism Quarterly*, *53*, 509–515.
Beniger, J. R., & Westney, D. E. (1981). Japanese and U.S. media: Graphics as a reflection of newspapers' social roles. *Journal of Communication*, *31*, 14–27.
Billings, R. S., Milburn, T. W., & Schaalman, M. L. (1980). A model of crisis perception: A theoretical and empirical analysis. *Administrative Science Quarterly*, *25*, 300–316.
Bitzer, L. (1968). The rhetorical situation. *Philosophy & Rhetoric*, *1*, 1–14.
Brown, H. (1987). *Society as text: Essays on rhetoric, reason, and reality*. Chicago: University of Chicago Press.
Burke, K. (1966). *Language as symbolic action: Essays on life, literature, method*. Berkeley: University of California Press.
Burke, K. (1969). *A grammar of motives*. Berkeley: University of California Press.
Charles, J., Shore, T., & Todd, K. (1979). *The New York Times* coverage of Equatorial and Lower Africa. *Journal of Communication*, *29*, 148–155.
Chinese Health Ministry releases information on encephalitis B outbreak. (2003, June 30). *Xinhua News Agency*, n.p. Retrieved January 21, 2005, from Lexis-Nexis.
Conway, M., & Hoskins, R. W. (2004, April 11). West Nile defenses raised; valley officials start early on campaign to keep mosquito-borne virus from spreading serious illness in this area. *Modesto Bee*, p. B1.
Coombs, W. T. (1999). *Ongoing crisis communication*. Thousand Oaks, CA: Sage.
Crable, R. E., & Vibbert, S. L. (1986). *Public relations as communication management*. Edina, MN: Bellwether.
Egbert, N., & Parrott, R. (2001). Self-efficacy and rural women's performance of breast and cervical cancer detection practices. *Journal of Health Communication*, *6*, 219–233.
Embler, W. (1966). *Metaphors and meaning*. Jacksonville, FL: Convention.
Encephalitis outbreak kills 24 in China, more than 300 infected. (2003, June 27). *The Associated Press*, n.p. Retrieved January 21, 2005, from Lexis-Nexis.
Farrell, T. B., & Goodnight, G. T. (1981). Accidental rhetoric: The root metaphors of Three Mile Island. *Communication Monographs*, *48*, 271–300.
Fink, S. (1986). *Crisis management: Planning for the inevitable*. New York: American Management Association.
Gates gives $27M grant for encephalitis. (2003, December 9). *Associated Press*, n.p. Retrieved January 21, 2005, from Lexis-Nexis.
Griffith, D. (2004, August 9). West Nile buzz often waved off. *Sacramento Bee*, p. A1.
Grissett, S. (2004, April 9). A few good hens drafted in the war on West Nile; blood samples track viruses. *The Times-Picayune*, p. A1.
Hall, C. T. (2004, April 12). West Nile makes early appearance in state; virus expected to move quickly from south to north. *San Francisco Chronicle*, p. A6.
Hathaway, M. (2004, September 2). County renews West Nile battle. *St. Louis Post-Dispatch*, p. 1.
Hayakawa, S. I. (1978). *Language in thought and action*. New York: Harcourt, Brace, Jovanovich.

Health officials say no evidence of encephalitis outbreak on Staten Island. (2003, October 12). *The Associated Press State & Local Wire*, n.p. Retrieved January 21, 2005, from Lexis-Nexis.

Hearit, K. M. (2006). *Crisis management by apology: Corporate response to allegations of wrongdoing*. Mahwah, NJ: Erlbaum.

Hearit, K. M., & Courtright, J. L. (2003). A social constructionist approach to crisis management: Allegations of sudden acceleration in the Audi 5000. *Communication Studies, 54*, 79–95.

Hopkins, T. S. (2003, May 22). Dona Ana County, N.M., combats mosquitoes, West Nile virus. *Las Cruces Sun-News*, n.p. Retrieved January 28, 2005, from Lexis-Nexis.

Ivie, R. L. (1987). Metaphor and the rhetorical invention of the Cold War "idealists." *Communication Monographs, 54*, 165–182.

Johnson, K. (2000, July 30). The things people choose to fear; usually they are unknown and uncontrolled, not near and dangerous. *The New York Times*, p. 27.

Kendall, J. E., & Kendall, K. E. (1993). Metaphors and methodologies: Living beyond the systems machine. *MIS Quarterly, 17*, 149–171.

King, W. (2004, June 15). Tracking the West Nile virus: Songbirds here are being tested for potentially deadly infection. *The Seattle Times*, p. A1.

Koch, S., & Deetz, S. (1981). Metaphor analysis of social reality in organizations. *Journal of Applied Communication Research, 9*, 1–15.

Kocieniewski, D. (2003, October 22). Victim of equine encephalitis is a 2-year-old, officials say. *The New York Times*, p. B5.

Lakoff, G., & Johnson, M. (1980). *Metaphors we live by*. Chicago: University of Chicago Press.

Lazar, K. (2004a, August 13). Virus has Bay State on bug alert; Brockton man in coma with EEE. *The Boston Herald*, p. 2.

Lazar, K. (2004b, August 14). Swat teams target skeeters; health officials take action to halt EEE virus. *The Boston Herald*, p. 5.

Ling, D. A. (1972). A pentadic analysis of senator Edward Kennedy's address to the people of Massachusetts July 25, 1969. *Central States Speech Journal, 21*, 81–86.

Lissack, M. (1997). Mind your metaphors: Lessons from complexity science. *Long Range Planning, 30*, 294–298.

Locke, C. (2004, August 29). Don't survey—deal with pest, officials say. *Sacramento Bee*, p. N5.

Mallan, C. (2004, June 25). Province slashes funding for West Nile. *Toronto Star*, p. B7.

McConnaughey, J. (2003, September 23). Three kinds of encephalitis virus now active in Louisiana. *The Associated Press*, n.p. Retrieved January 21, 2005, from Lexis-Nexis.

Millar, F. E., & Beck, D. B. (2004). Metaphors of crisis. In D. P. Millar & R. L. Heath (Eds.), *Responding to crisis: A rhetorical approach to crisis communication* (pp. 153–166). Mahwah, NJ: Erlbaum.

Orr, C. J. (1978). How shall we say: "Reality is socially constructed through communication?" *Central States Speech Journal, 29*, 263–274.

Over 160 die of encephalitis in India. (2003, July 23). Xinhua News Agency, n.p. Retrieved January 21, 2005, from Lexis-Nexis.

Perez-Pena, R. (2003, September 21). Who's afraid of this little fellow?; West Nile is still here, but the furor and anxiety have ebbed. *The New York Times*, p. 41.

Propp, K. M. (2003). Metaphors of survival: A textual analysis of the decision-making strategies of the *Survivor* contestants. In M. J. Smith & A. Wood (Eds.), *Survivor lessons: Communication issues under a watchful eye* (pp. 111–131). Jefferson, NC: McFarland & Company.

Purdy, M. (2000, July 23). Public health meets public relations in the battle of the bugs. *The New York Times*, p. 21.

Reynolds, B. (2002). *Crisis and emergency risk communication*. Atlanta: Centers for Disease Control and Prevention.

Reynolds, B., & Seeger, M. W. (2005). Crisis and emergency risk communication as an integrative model. *Journal of Health Communication, 10*, 43–55.

Seeger, M. W., Sellnow, T. L., & Ulmer, R. R. (1998). Communication, organization, and crisis. In B. R. Burleson (Ed.), *Communication yearbook 21* (pp. 231–275). Thousand Oaks, CA: Sage.

Seeger, M. W., Sellnow, T. L., & Ulmer, R. R. (2003). *Communication and organizational crisis*. Westport, CT: Praeger.

Simhan, R. (2004, August 24). Fears and facts most people worry about high-profile hazards but ignored common—and often deadlier—dangers. *Sacramento Bee*, p. E1.

Singleton, D. (2003, October 12). Five fight encephalitis, skeeter virus ruled out. *New York Daily News*, p. 5.

Station, W. (2004, October 6). Statement of Wendy Station Founder Encephalitis Global. *Congressional Quarterly*, n.p. Retrieved January 21, 2005, from Lexis-Nexis.

Theus, K. T. (2004). Issue management during sudden executive departures: Sensemaking, enactment and communication. In D. P. Millar & R. L. Heath (Eds.), *Responding to crisis: A rhetorical approach to crisis communication* (pp. 327–346). Mahwah, NJ: Erlbaum.

Viral encephalitis: Outbreak reported in southern China. (2003, July 15). *Virus Weekly*, p. 20.

Weick, K. E. (1979). *The social psychology of organization* (2nd ed.). Reading, MA: Addison-Wesley.

Weick, K. E. (1988). Enacted sensemaking in a crisis situation. *Journal of Management Studies, 25*, 305–317.

Weiner, M. (2004, September 19). Experts don't know why EEE is back; little research on deadly illness. Doctors wonder why more not sick. *The Post-Standard*, p. A1.

Werde, B. (2003, September 30). 3 Suffolk parks are closed nightly after a deadly virus is detected. *The New York Times*, p. B5.

Witte, K., Meyer, G., & Martel, D. (2000). *Effective health risk messages*. Thousand Oaks, CA: Sage.

10

ORGANIZING MULTIPLE HIV AND AIDS DISCOURSES FOR ENGAGING CRISIS COMMUNICATION IN INDIA

Avinash Thombre

India has overtaken South Africa recently as a country with the largest HIV positive population. The present analysis uses Parrott's (2004) framework of identification of discourses in the HIV/AIDS context in India to help effectively design risk and crisis interventions. HIV/AIDS discourses in India are identified in the *lay, societal,* and *expert* arenas. This chapter examines stigmatization of HIV and AIDS as a factor in India's inability to heed warnings to avert crisis situation. Specifically, the present chapter explicates discourses that allow HIV and AIDS risk in India to incubate and to erupt into a future public health crisis of a significant magnitude. Coherence of discourses based on a continuum of communication and health care is suggested.

It [HIV AND AIDS] will slice through India like a hot knife through butter.

—Bill Clinton, former president of the United States,
at the Global Business Council on HIV/AIDS
Awards Dinner, New York, June 2002

India is projected to play a key role in future global affairs. There has been extensive media coverage and speculation related to its HIV and AIDS disease burden. HIV infections in the past 5 years have increased substantially, nearly bringing the country closer to a health crisis. Current official estimates indicate that there are 5.134 million HIV-infected people in India constituting about 6% of the total population (National AIDS Control Organization, 2005; "You and AIDS," 2004). Although the percentage is small, the numbers are staggering, the highest of any country, overtaking South Africa (UNAIDS, 2004). Understanding HIV/AIDS in India as a public health problem is complex. The present analysis focuses on stigma as one of the primary contributing factors to the HIV and AIDS near crisis situation in India.

Crises typically are understood as a broad class of events that are disruptive, unanticipated, and threatening (Seeger, Sellnow, & Ulmer, 2003). In particular, public health crises are problematic and challenging because most have negative outcomes associated with loss of human life. Traditionally, crisis communication begins after risks emerge or erupt into a crisis on a large scale.

The complex process of sense-making of a crisis event was examined by Turner (1976) through a six-stage failure-of-foresight model. The first stage of Turner's conceptualization is a period in which subtle changes in environment go unnoticed. For example, a few HIV infections in the early 1980s in Africa were hardly considered abnormal or threatening. In the second incubation stage, an accumulation of events leading toward a crisis occurred. In the case of Africa's HIV and AIDS crisis, the incubation period was sudden, a matter of a few years, eventually leading to a full-fledged crisis.

The first signs of a crisis are manifested in the third precipitation stage, the period that signals the eruption of threat into a crisis. For example, at this stage, there is an actual loss of life or damage to property, creating chaotic situations. It is at this stage that process becomes distinctly visible. However, the magnitude of the crisis has not yet reached the catastrophic phase. In the fourth stage, called the *onset*, the crisis is full blown, causing extensive damage and large-scale disruption to normalcy. This stage is characterized by long periods of loss of life. The last two stages involve rescue and salvage and a series of cultural adjustments that call for acceptance of changes in norms and circumstances.

The current situation in India of stigmatization of those diagnosed with HIV/AIDS infections is leading to a failure to heed warning signals and corresponds to Turner's second incubation stage. Stigmatization causes people and authorities to fail to see that a risk is real, leading them to ignore it and thereby allow it to attain crisis status. With continued failure to recognize warning signals, the situation has the potential to quickly develop into the precipitation stage and, further, into the onset stage. It results in large-scale deaths in vulnerable populations.

Recent conceptualizations of crisis communication by Seeger, Sellnow, and Ulmer (2003) suggest risk communication as part of the preeruption or precrisis stages. Defined by a variety of groups, risk communication differentiates from crisis communication as any two-way exchange between stakeholders about the existence, nature, form, severity, or acceptability of risks. It addresses factors leading to likelihood of occurrence of a particular crisis and tries to predict its magnitude. Based on these factors, it advocates allocation of resources for research and control (Covello, 1992). Thus, risk and crisis communication go hand in hand and focus on the processes of information and opinion exchange with the public (National Research Council, 1989). Efforts to communicate in public health crisis situations, however, are hampered by stigma known as an interactional process within societies in which particular social identities are collectively devalued (Ogden & Nyblade, 2005).

To address stigmatization of those infected with HIV and AIDS to stop or slow the rate of infections in the context of public health crises, examination, and understanding of variety of discourses is essential. The HIV and AIDS discourses have a remarkable ability to generate meanings, which is apparent when one examines the multiplicity of its interpretations (Gilmore & Somerville, 1994; Pittam & Gallois, 2000; Sontag, 1991). HIV/AIDS is referred to as a punishment from God or a *gay plague*. In Africa, traditional healers describe it as an *eating* away at the body, as a *greedy* activity of the virus—a thing that sucks life out (Garrett, 2000; Wolf, 2002). In India, HIV/AIDS is referred to as a disease of new untouchables (Singhal & Vasanti, 2005).

Parrott (2004) advocates a framework of identification of these main discourses under the larger domain of health communication and public risk communication in order to effectively understand risk interpretations and design interventions. Identification of *lay discourse* focuses on understanding how the general public frames a particular disease. Lay discourses constitute use of indigenous knowledge sources based on cultural, social, and individual practices to guide individual health behavior and health care outcomes. A next-level discourse, *societal discourse,* concerns allocation of scarce resources, including money for scientific and behavioral research and health care delivery services. It focuses on time spent analyzing, discussing a health problem, and setting an agenda in the political, religious, and organizational arenas to champion health delivery and services.

The final discourse, called *expert discourse*, concerns the understanding and use of expert scientific information and knowledge about health and health care in order to inform, motivate, and set rules for profit-making. Parrott (2004) emphasized that identifying these discursive practices can help develop a unified preventive communication strategy inclusive of the many discourses on the issue. It would mean not ascribing attention to select discourses at the expense of ignoring other discourses in an effort to design coherent risk prevention strategies.

Based on the previous framework, the purpose of present chapter is to (a) understand and explicate HIV and AIDS risk discourses in the societal, expert, and lay spheres in India; and (b) examine stigmatization of HIV and AIDS as a factor in India's inability to heed warnings to avert crisis situation. Specifically, the present chapter explicates discourses that allow HIV and AIDS risk in India to incubate and to erupt into a future public health crisis of a significant magnitude. The analysis is based on random and informal open-ended discussions conducted in late 2004 with HIV and AIDS experts, HIV-positive individuals, and health care workers in Pune, India. Data for the analysis are also based on archival materials, newspaper clippings, and analysis of Web-based discourses. Implications for risk and crisis communication are discussed.

LAY HIV AND AIDS DISCOURSES IN INDIA

As discussed in the introductory section, the foci of lay health and health care discourse centers on the use of indigenous culture-based knowledge sources. The discourse relies on experiential information derived from social, cultural, and individual arenas that guide individual behavior with health and health care outcomes (Dutta-Bergman, 2004). Specific HIV- and AIDS-related cultural beliefs further intersect practices, social networks, norms, and the role of perceptions about expectancies within one's social groups. Understanding lay discourses is crucial because these may overshadow rational decision making, hampering effective risk and crisis communication. The knowledge of these discourses helps frame appropriate design of more ecologically sound message interventions (Parrott, 2004).

HIV and AIDS lay discourses among the majority of low, middle, and upper class Indians are broadly associated with three components: (a) denial of being HIV and AIDS positive, (b) fear of HIV and AIDS, and (c) discrimination and stigmatization as a result of HIV and AIDS. These lay discourses constitute multiple sublayers. Identifying the many sublayers is imperative to designing effective risk and crisis communication strategies. Although denial is connected to fear of HIV and AIDS, stigmatization and discrimination can be manifested in overt or covert forms. Nelson Mandela (2000) asserted that many HIV- and AIDS-infected individuals are killed not by the disease, but the stigma surrounding the disease. This assertion holds true for most cases in India. Using Parrott's framework, the following paragraphs illustrate the manifestation of HIV and AIDS denial, fear, and discrimination discourses in the lay sphere in India.

Denial Discourse

The first Indian case of HIV infection was detected in 1986 in Chennai (Solomon, Chakraborty, & Yepthami, 2003). The progressive development of present lay HIV and AIDS stigmatization discourse has its roots in the initial denial phase of early HIV infections in India. In the first 10 years, a distinct HIV or AIDS discourse in the public arena was absent due to scant knowledge and discussion (www.unaids.org; Singhal & Rogers, 2003). The social taboo of discussion of sexual matters further prohibited frank discussion of the causes and spread of HIV. This led to the creation of a denial discourse among the lay population. Many did not know that they suffered from HIV and AIDS, which led others to deny HIV or simply ignore it.

As the years progressed and statistics increased, the infected were vaguely referred to as suffering from tuberculosis (an opportunistic infection), not HIV or AIDS. The social taboo increased in recent years, resulting in a distinct absence of naming HIV and AIDS infections in the lay discourse. Despite several government-sponsored information campaigns, common people even today acknowledge their relatives infected with HIV and suffering from AIDS in a discreet manner. The infected are referred to as simply *ill*.

If one persists in knowing the reason for being ill, most answers describe that the cause is unknown. In common parlance, HIV is referred to as some kind of bad disease. Any further discussion of the exact nature of suffering is discouraged and avoided. In general, the silence and rumors associated with HIV and AIDS result in the spread of fear and lack of knowledge of transmission routes (UNAIDS, 2002). Hence, early denial discourses resulted in the creation of an atmosphere ripe for the next level of fear and discrimination discourses.

Fear Discourses

Prominent HIV and AIDS lay discourses at the individual level consist of fear of ill treatment. HIV-positive people are afraid of various illnesses, debilitating ill health, painful conditions, and social rejection. The discourses center on cases of medical neglect, such as individuals being denied admission to hospital or desertion during treatment. The discussions focus on loss of significant relationships, trust, and confidence (UNAIDS, 2002). Lay discourses frame HIV and AIDS as associated with, or resulting in, loss of job or income and fear of damaging or losing family reputation.

Next, fear discourses center on social ostracizing and isolation—of being avoided or shunned by close family members. Death, dying early, dying uncared for, being denied last rites, and social ridicule form a central part of lay fear discourses. A sublayer of fear discourses is the common fear of being identified with deviant, morally sinful behavior, mainly sexual promiscuity and visiting sex workers. Lay fear discourses in work settings are expressed in covert forms. Many employers terminate or refuse employment to HIV-positive people out of fear of transmitting infections to other employees.

In other contexts, fear-related interpersonal discourses can be examined among health care staff. Because a large section of the middle and lower level health care staff lacks knowledge about medically appropriate ways to treat positive people without themselves getting infected, incidents of medical staff being infected accidentally by coming in contact with positive people circulate. This lack of knowledge translates into fear of HIV- and AIDS-infected people, leading to ill treatment of positive people.

The ill treatment is manifested in refusal to provide treatment for HIV- and AIDS-related illness and refusal to operate or assist in clinical procedures (Bharat, Aggeleton, & Tyrer, 2001). For example, HIV-positive people are provided restricted access to facilities such as toilets and common eating and drinking utensils. Many patients are confined or isolated in separate wards or given a separate bed outside the ward in a gallery or corridor. Cases of some being denied ongoing treatment are common.

Fear discourses increase in severity when health staff use protective gear (gowns, masks, etc.) when treating positive people. Cases of health staff refusing to lift or touch a HIV-positive person's dead body and using plastic sheeting to wrap the body are common. There are reported delays in treatment, including slow service and a reluctance to provide transport for the body. Positive people are made to wait in queues or asked to come again. These discourses form a central part of fear of HIV among lay people (Bharat et al., 2001). The fears circulate, reinforcing misconceptions of HIV and AIDS and resulting in continued fear which further led to discrimination and stigmatization.

Discrimination and Stigmatization Discourses

Discrimination is at the heart of lay discourses. It is especially directed against women identified as HIV-positive. A common occurrence of discrimination is rejection of positive women by family members. Women are accused of not controlling their partners' urges to have sex with other women. In some cases, women whose husbands have died from AIDS-related infections are blamed for their deaths. For example, Kareena, an HIV-positive woman, said, "My mother-in-law tells everybody, Because of her, my son got this disease. My son is simple as good as gold—but she brought him this disease." In home settings, infected women are highly stigmatized and often find themselves discriminated by their family members. Sunita, an HIV-positive woman, for example, narrated her experience,

> My mother-in-law has kept everything separate for me—my glass, my plate, they never discriminated like this with their son. For me, it's don't do this or don't touch that and even if I use a bucket to bathe, they yell—"wash it, wash it." They really harass me. I wish nobody comes to be in my situation and I wish no body does this to anybody. But what can I do? My parents and brother also do not want me back.

Common instances of discrimination are refusal to share property or access to finance. Women are blocked access to spouse, children, or other relatives and even subjected to physical isolation at home. Blocked entry to common areas or facilities such as toilet or neighborhood areas is a common occurrence. Children of HIV-positive parents are often denied the right to attend school or are segregated from other children. The manifestation of discourses is outlined in Table 10.1.

People living with HIV/AIDS who choose to disclose their status at work experience extreme discrimination, including removal from job or forceful resignation. Many are punished by withdrawal of health/insurance benefits and provided restricted access to shared facilities. Labeling and name-calling occur frequently (Bharat et al., 2001). Mohan, an HIV-positive man, age 27, said, "Nobody will come near me, eat with me in the canteen, nobody will want to work with me, I am an outcast here."

In its overt forms, stigmatization discourses lead to the ill treatment of bodies of those who die of AIDS-related illnesses. In Mumbai, bodies of HIV/AIDS victims are placed purposely in black plastic bags, a practice that makes finding a good undertaker and funeral services difficult. In several cases, infected bodies are not administered traditional burial rites, a serious loss of social prestige in a traditional society. In Kerala, the highest literate state in India, bodies of HIV and AIDS victims were refused the most basic right of burial rites or space in the cemetery, a case of acute stigmatization (Raghavan, 2005).

In its most visible overt forms of lay stigmatization, discourses are manifested in extreme violence toward HIV-infected people. The most violent form of stigmatization could mean death for individuals living with HIV. For instance, an HIV-positive woman in Dharsana village in the Dhandi district of Gujarat State in India was ostracized, beaten, and gruesomely murdered. Villagers stoned her because she carried the infection (Thomas, 2005). A common justification offered as part of lay stigmatization discourses is that the woman deserved her fate when she acquired this disease.

In contrast to lay discourses, the societal discourses differ in their meaning and naming of HIV. Using Parrott's organizing framework, the following section explicates the present HIV- and AIDS-related societal discourses in India.

TABLE 10.1.
Manifestation of HIV and AIDS Lay Discourses in India

LAYER OF LAY DISCOURSE	FORMS OF MANIFESTATION
1. Denial	1. Inhibition of frank discussion about HIV and AIDS
	2. Refusal to learn HIV transmission routes
	3. Silence
	4. Rumors about HIV and AIDS
2. Fear	1. Suffering associated with having infection
	2. Refusal to hospital treatment
	3. Loss of relationships, social networks
	4. Loss of reputation, power, standing in society
	5. Isolation, public rejection
	6. Being labeled deviant
	7. Loss of customer/business
	8. Blamed for bad luck of family
	9. Loss of marriage, childbearing opportunities
	10. Physical violence (beatings, throwing stones, arrests)
	11. Confinement/restriction to house or a particular space
	12. Separation from children
	13. Threats, mocking, insults
3. Discrimination	1. Harsh treatment by community
	2. Losing income, property
	3. Losing face
	4. Fair treatment in employment opportunities
	5. Differential treatment in public spaces, health settings, schools
	6. Denial of housing, eviction by landlord
	7. Loss of employment
	8. Denial of loans, scholarships
	9. Provision of substandard health treatment
	10. Refusal of services (food, facilities, etc.)

Note. Adapted from Ogden, J., & Nyblade, L. (2005). *Common at its core: HIV-related stigma across contexts*. Washington, DC: International Center for Research on Women.

HIV AND AIDS SOCIETAL DISCOURSES IN INDIA

The existence of societal discourses can be examined when individuals in influential positions engage in a meaning-making process of HIV/AIDS to frame policies, laws, and their implementation (Parrott, 2004). The meaning-making discourses set the tone for deciding priorities for health research and policy for HIV and AIDS. Collective societal discourses help shape future scientific knowledge generated in a society. It also determines in what form the knowledge will be available and the kind of evidence that will be used to promote public health. Emergent knowledge generated from societal discourses is, in a critical way, important in shaping future crisis and risk communication and impact intervention strategies. As in the case of lay discourses, manifestation of societal discourses can be examined as cases of denial, fear, and discrimination of HIV and AIDS individuals. The various manifestations of societal discourses are outlined in Table 10.2.

TABLE 10.2.
Manifestation of HIV and AIDS Societal Discourses in India

LAYER OF SOCIETAL DISCOURSE	FORMS OF MANIFESTATION
1. Dismissal	1. Lack of acceptance of the magnitude of HIV and AIDS as public health problem 2. Meager allocation of resources 3. Not acknowledging number of real deaths due to HIV 4. Turning a blind eye to transmission routes
2. Fear	1. Cultural repercussions 2. Political fallout 3. Loss of power 4. Religious fallout
3. Discrimination	1. Shunning certain populations—sex workers, homosexuals, drug users 2. Branding deviant groups dangerous 3. Finger pointing at other officials 4. Human rights abuses

Denial and Dismissal Discourses

A substantial segment of HIV and AIDS discourses among decision makers, particularly among the governmental officials in India, relate to denial and dismissal of the existence of HIV/AIDS (Singhal & Rogers, 2003). During the early 1990s, the federal health minister and health officials either extremely downplayed the presence of HIV and AIDS or denied it outright. In cases where the presence of infections was acknowledged, it was portrayed as having no impact on India, and therefore a minimal cause of concern. Societal discourses primarily framed HIV as a Western or African problem that does not have a substantial effect on India (Singhal, 2004). India's strong cultural values with regard to sexuality were cited as helpful and sufficient to protect it from the rapid spread of HIV and AIDS infections.

Global Fund to Fight AIDS Executive Director Richard Feachem (2003) noticed the denial and commented:

> There is a fairly widespread view among educated people and opinion leaders in India that HIV and AIDS is primarily an African problem and that Hindu and Muslim culture will protect India from the most serious consequences of the virus. . . . As in other countries, there has been a resort to the mythology of cultural immunity. It can't happen to us because we're different. . . . I found on my visit a persistent tendency to minimize the current scale of the epidemic and the potential future growth.

The denial discourse increases because there exists a lack of consensus within the political establishment and among nongovernmental organizations (NGOs) about the extent of the HIV pandemic and the right strategy to combat it. The other bone of contention relates to the method of dealing frankly with sexuality (Chatterjee, 2003). For example, former Chief Minister of Andhra Pradesh Chandrababu Naidu favored a direct approach to dealing with sexuality and the spread of HIV. He wanted condoms to be displayed and promoted at official functions. He initiated a project to hand out free condoms at liquor shops, making them available through vending machines at cafes on highways and at railway stations. However, federal Health Minister, Sushma Swaraj, opposed the idea, fearing it would go against existing conservative sexual norms. She reasoned that promoting condom use would encourage immoral sexual behavior. These instances of a lack of consensus hamper risk prevention efforts.

Societal discourses by individuals in positions of power are responsible for continued denial and inaction (Chatterjee, 2003):

> The problems facing HIV and AIDS prevention efforts are compounded by the refusal of many Indians in positions of power to accept that their country faces a grave threat from the pandemic. On the surface, HIV and AIDS seem to have become a priority. As long ago as 1998, Prime Minister Atal Behari Vajpayee conceded, HIV and AIDS is the most serious public health problem facing India. And AIDS is now a fashionable cause among actors, fashion divas, rock stars, and socialites. But when the discussion of AIDS shifts from the vague to the explicit, from talk to action, problems crop up. (p. 1526)

The effect of dismissal and denial discourses are reflected in the meager allocation of resources to combat the spread of HIV. A total of $38.8 million of the government's own funds were allocated for a 5-year period (1999–2004) to tackle HIV/AIDS, demonstrating a lack of serious commitment (Avert India, 2005). Despite the obvious increase in infection rates, denial discourse of people in positions of power has not changed, and lessons have not been learned. This is reflected in the so-called enhanced outlay of a meager $11.08 million for the entire health department for 2004. Of the total outlay, only $2.34 million has been allocated for HIV and AIDS prevention and treatment for 2004–2005 (Ekstrand, Garbus, & Marseille, 2004). Thus, denial discourses continue to hamper prevention efforts.

Fear Discourses

Government, media, and research officials responsible for HIV federal and state prevention strategies and media campaigns unknowingly help create fear of certain sections of the population. Risk prevention messages designed and developed by officials target the role of women engaging in commercial sex work. The association of HIV with commercial sex workers is interpreted by the general population as a disease that is affecting only immoral people or those associated with immoral behavior such as prostitution.

In the larger society, messages targeted toward commercial sex workers create a backlash. Several media reports describe violence against female sex workers, creating fear of HIV and AIDS and targeted interventions.

Discrimination and Stigmatization Discourses

Although discrimination and stigmatization of HIV-infected individuals is rampant, there is a lack of mainstream societal discourse on it. The discrimination cases are hushed up by officials, portrayed as only isolated incidents and not as a norm. Wishful thinking on the part of officials exists. There is a belief that if HIV discrimination is not talked about out loud, it will disappear on its own. Cases of discrimination of daughter-in-laws—those in workplace and medical settings—are ignored. Human rights abuses are rampant, including discrimination perpetuated by government officials and employees directed at people living with HIV/AIDS, sex workers, and male's having sex with males, as well as volunteers who work on HIV and AIDS projects (S. Pujari, personal communication, 2004).

There is no explicit national policy on HIV and AIDS that addresses discrimination and no national legislation to serve as an implementing instrument. It is rare for those participating and perpetuating discrimination to face punishment. Cases of violence against HIV-positive individuals in rural areas rarely result in punishment. However, a few lawyers in Mumbai and New Delhi have recently started providing voluntary legal services free of charge, thus incriminating some responsible for discrimination.

Finally, sense-making associated with HIV and AIDS differs substantially in the realm of expert discourses from the societal discourses. The following section explicates meaning-making of discourses associated with HIV and AIDS in the expert arena.

EXPERT HIV AND AIDS DISCOURSES IN INDIA

Understanding expert discourses is fundamental to risk prevention. These discourses are developed by experts in the field and focus on use of expert bioscientific information and knowledge about health and health care related activities. Expert discourses primarily shape the methodologies and strategies used in informing the general public and are important in risk and crisis prevention. These discourses set the tone for formulating policies of motivating the direction of health research (Parrott, 2004). The method by which expert discourses are constructed, interpreted, and disseminated (particularly with regard to risk construction about HIV and AIDS and its transmission, prevention, and coping mechanisms) influences whether these messages are accepted or rejected by lay people.

Medical doctors, scientists, and prevention experts engage, formulate, and shape expert discourses of HIV and AIDS in India. In addition, communication experts and individuals trained in handling public relations also actively shape these expert discourses. These individuals are highly trained and competent, and they set the agenda for designing the direction of future prevention research, allocation of monies to care centers, communication activities, and prevention strategies. The following paragraphs illustrate the presence of expert discourses in the realm of fear, denial, and discrimination areas (Table 10.3).

TABLE 10.3.
Manifestation of HIV and AIDS Expert Discourses in India

LAYER OF EXPERT DISCOURSE	FORMS OF MANIFESTATION
1. Fear	1. Beware of commercial sex workers 2. Beware of homosexuals
2. Denial	1. Not adequate use of condoms 2. Rampant existence of multiple partners 3. Acknowledgment that HIV has spread to general population
3. Discrimination	1. Blaming certain groups as problematic 2. Blame-oriented media campaigns

Fear-Based Discourses

A closer examination of present discourses in expert HIV and AIDS prevention in India reveals its organization into fear-based risk-construction messages (Singhal & Rogers, 2003). Primarily, HIV expert discourses in India classified HIV and AIDS risk as a demographic variable, rather than a behavioral factor. Based on epidemiological and public health disease transmission data, experts identified risk groups as men having sex with men, hemophiliacs, heroine addicts, and commercial sex workers.

Prevention messages targeting these specific populations were disseminated on a large scale throughout the country. After identifying certain practices (not using condoms, engaging in multiple partners or risky sexual behavior), these messages asked the larger population to stay away from risky populations and their practices. Intense media campaigns further perpetuated fear of particular groups and created a notion that the general population is not at risk.

Prevention efforts designed to prompt individuals into recognizing the heightened probability of HIV infection linked to these and other activities thus create a cycle of fear, leading to denial and discrimination (Estrada & Quintero, 1999; Fisbein, 2001). Next, experts and their subordinates were in denial for a long time about the existence of HIV cases.

Denial Discourses

Similar to discourses engaged by lay people and those in positions of power, the knowledgeable experts too were in self-denial of the real magnitude and extent of transmission of HIV and AIDS in the general population. Due to lack of accurate data-collection methods, experts were in denial of the rampant existence of multiple partners. Acknowledging that infections jumped from risky populations into the general population was another aspect of serious denial.

As a result, experts failed to seriously take into account the rapid spread of HIV and AIDS. One of the negative consequences of denial discourses was the inadequate distribution and emphasis on use of condoms to the targeted risky population. The impact of denial discourses meant that research on the transmission patterns and identification of different strains of HIV virus in the Indian population was carried out by the scientific community with remarkable sluggishness. Likewise, the communication campaigns launched by many governmental and private agencies were not targeted to the general population nor given enough air time.

Discrimination Discourses

The critical association of stigmatization and discrimination with HIV and AIDS in the general population was ignored by experts during the 1990s. The discrimination discourses engaged by experts were mostly during the early phases of HIV infections (late 1980s and early 1990s). Instances of the presence of discrimination discourses found among experts are primarily cases of unconscious blame in the media campaigns. For example, communications strategies are based on "don't do this because this will happen." Prevention messages were blame-oriented and advised the larger population not to hang around with those with alternative sexual lifestyles. As a result of these messages, the affected people were feared, shunned, and ostracized by society (Singhal, 2004). Cases of discrimination were thus unknowingly perpetuated. In some places, expert discourses continue to be blame-oriented, moralistic, preachy, and didactic.

The final section of the present chapter advocates a strategy to coherently combine lay, societal, and expert discourses to draw a meaningful framework for tackling stigmatization associated with HIV and AIDS in India. The idea is to engage specific discourses with each other and create an atmosphere of openness that strengthens acceptance of HIV-positive people. The aim of engaging the three different discourses is to help influence and design more coherent HIV prevention messages. The coherent message design strategy will hopefully help contain the present crisis situation in India from reaching a full-blown crisis like Africa.

Engaging Discourses: Coherence in Construction of HIV and AIDS Prevention Messages

> The choice now is clear and stark: India can either be the home of the world's largest and most devastating AIDS epidemic—or it can become the best example of how this virus can be defeated. (Bill Gates, founder, chairman and chief software architect, Microsoft Corporation)

There are apparent differences in sense-making of HIV and AIDS risk construction in India in the lay, societal, and expert discourses. Understanding the differences in these discourses is central to understanding the incongruence in meaning-making and confusion in risk message interpretation by the general population. The present analysis clearly demonstrates presence of denial, fear, and stigmatization in lay, societal, and expert domains. The complex cultural and social meanings given to prevention messages by lay people resulted primarily in creating fear, denial, and discrimination. Likewise, the different meaning ascribed to HIV/AIDS by people in positions of power and experts, is problematic to developing effective strategies.

Analysis of successes and failures of prevention discourses hampered by stigmatization of those diagnosed with HIV and AIDS reveals problems at multiple levels. In the lay sphere, there is a distinct presence of denial, fear, and discrimination discourses that are independently interpreted by the general population. These separate discourses result in complete failure of prevention messages. A direct result of not engaging lay discourses is a continued increase in the number of infections. The failure leads to misconceptions and increased stigmatization of HIV and AIDS.

In the expert sphere, presence of unintentional denial discourse is evident, whereas presence of denial discourses in the societal sphere is minimal. An area of concern is the presence of denial, fear, and discrimination discourses, in substantial levels in the lay sphere. Fear discourses exist in substantial levels in societal and expert discourses leading to large-scale incongruity in interpretation of prevention messages. Table 10.4 explains the levels of these discourses in the three spheres.

Recognition of the existence of multiple layers of lay, societal, and expert discourses is a first step in understanding how HIV is understood in India. The present chapter advocates that effective and efficient HIV and AIDS risk prevention in India is possible by acknowledging denial, discrimination, and stigmatization associated with HIV and AIDS. It is important to engage the different levels of present expert, societal, and lay discourses with each other to arrive at a coherent, integrative framework. Coherence in discourses helps develop appropriate risk prevention messages that are engaging, involving, and enlightening (Parrott, 2004). A strategy of coherent care-based discourses is proposed in the following paragraphs.

TABLE 10.4.
Levels of Presence of HIV and AIDS Communication Discourses in India in Three Spheres

DISCOURSES/SPHERES	LAY	SOCIETAL	EXPERT
Denial	Present (substantial)	Present (minimal)	Present (substantial)
Fear	Present (substantial)	Present (minimal)	Present (substantial)
Discrimination	Present (substantial)	Present (minimal)	Present (minimal)

Care-Based Coherent Discourses

The proposed coherent discourse is based on creating a model of comprehensive dialogue-based prevention programs that are backed by holistic care, treatment, and support services. Based on a continuum of care for people living with AIDS, the coherent discourse approach borrows aspects of the comprehensive Y. R. Gaitonde model presently implemented on a trial basis in south India (YRG Care, 2005). The communication and care-centered model involves transfer of core values concerning patient-centered care, client confidentiality, and the broader involvement of families and communities, keeping in mind different contexts, cultures, and resources.

HIV infections in India jumped from high-risk populations to the general population recently. To stem the rate of infections, as a first step in implementation of the communication and care-centered model, a change in expert discourse needs to take place. It is suggested that the expert discourse focus on improving clinical, laboratory, and technical capacities of healthcare staff. Next, the expert and societal discourses can focus on orienting the general population and the HIV- and AIDS-positive people to think of themselves as normal and seek help in coping with the disease. A similar experiment in a patient-based care model is already being implemented in south India with encouraging success (YRG Care, 2005).

Societal discourses can focus on campaigns that provide and enhance support facilities such as counseling and social support services. Officials and policymakers should advocate high-quality voluntary counseling and testing services and make sure that it is provided in many areas. Evidence suggests that the introduction of testing and treatment in affected communities reduces the fear, stigma, and discrimination surrounding HIV and AIDS. It also increases demand of HIV testing and counseling and reinforces prevention efforts (WHO, 2004). Together, these efforts further reduce stigmatization.

Framing coherent messages in expert and societal discourses drives home the message that lax norms of male sexual behavior are dangerous for men, their wives, and their families. Further, encouragement provided to develop messages documenting the advantages of being aware of an individual's HIV status helps reinforce beliefs that one can proactively protect one's health and live a longer and happier life. Examples highlighting that someone dear to any one of us could be HIV positive helps personalize the disease and wring out associated stigma.

At the core of engaging expert discourses is recognition of the prevalence in lay discourses of a genuine fear of people living with HIV and AIDS. Expert and societal discourses can heed this genuine fear of the disease by acknowledging it is important and natural to fear HIV and AIDS. However, highlighting positive examples of individuals who have successfully lived with the virus as productive members of the society in a mass media campaign can help change lay fear and stigmatization discourses. The campaign results in reducing fear of people living with HIV and AIDS, creating an atmosphere of acceptance and understanding in the majority population.

Discourses in the expert realm can focus on generating beliefs stressing the importance of continuing normal relationships with an HIV-positive person. This convinces individuals engaging in lay fear discourse that HIV can be managed effectively. It also helps reduce stigmatization and discrimination. Emphasis of culturally appropriate examples of HIV-positive people living longer and productive lives with healthy lifestyles, extending and giving love and support would help. Messages designed to respect and support HIV-positive people willing to declare their status creates a social environment that allows them to do so and further neutralize stigmatization.

Steps to engage societal discourse among officials and politicians on HIV and AIDS to monitor education programs that help increase health awareness in the community need to be a national priority. This increased health awareness results in diminished stigmatization of the diseased members of society. Interactive and integrative health communication-based education programs such as media campaigns should address issues of sex and sexuality that traditionally are not discussed openly in India.

Action-based societal discourse can follow by preventing socially vulnerable groups from becoming infected with HIV. Strict guidelines to protect the rights of people living with HIV/AIDS to education, work, and privacy, and to ensure care and support, need to be put in place and implemented strictly. Officials can combat the discrimination of HIV-positive individuals by healthcare workers by providing adequate training in clinical management of HIV and AIDS.

Further, an official announcement of India's National AIDS Prevention and Control policy at the federal level can reduce confusion and provide clear direction to officials. A movement to start HIV and AIDS patient-friendly hospitals needs to be strengthened (WHO, 2004). All government hospitals should be instructed to admit HIV and AIDS cases without discrimination as a policy. Discharge certificates should only refer to their opportunistic infections, without mention of HIV seropositivity, although the case sheet at the hospital keeps those records.

Concerted effort at the political level forms a large part of coherence in discourses. A definite positive start in this area is already evident. For example, former Prime Minister Vajpayee addressed Indian parliament, acknowledging HIV and AIDS as one of the most serious health challenges facing the country. Present Prime Minister Man Mohan Singh followed this line and expressed serious concern about the spread of HIV and AIDS, arguing to intensify prevention activities. In addition, he met business and industrial community members interested in addressing the HIV/AIDS epidemic. These efforts point to coherence in societal discourses. Such leadership initiatives need to be followed down official chains of action.

Media needs to be an integral part of the overall coherent discourse. Responsible media messages highlighting HIV and AIDS individuals who want to come out in the open and show they are living successful lives are essential. Prosocial entertainment education programs to change and challenge existing notions of fear, denial, and discrimination discourses on HIV and AIDS in India need to be a priority (Singhal & Vasanti, 2005). For example, an engaging story of a likeable heroine who, with her son's support, takes a courageous stand to acknowledge her positive status can start a new debate.

Social learning from these engaging societal discourse strategies should be encouraged. For example, in Thailand, the minister for tourism, public information, and mass communication, Mechai Viravaidya, appeared on national TV and took a sip of Coca-Cola from the same glass as an HIV-positive child. This strategy had a tremendously destigmatizing influence on the Thais. India needs to create its own Viravaidya, Rock Hudson, or Magic Johnson to neutralize and engage the lay stigmatization discourses (Singhal & Rogers, 2003).

Finally, possible channels for communicating risk and overcoming stigma of HIV and AIDS include traditional media such as puppet shows, music, and drama. These channels need to be used in conjunction with printed educational support materials on HIV/AIDS, one-to-one interpersonal communication, peer education, and discourses engaged by NGOs.

CONCLUSION

Disengaging HIV and AIDS discourses in India related to fear, denial, and stigmatization acts as a powerful barrier to the process of crisis and risk message interpretations. Any mechanism to reduce or engage these negative discourses needs to be embraced. A shift in developing crisis and risk communication messages that harmoniously engage multiple HIV and AIDS discourses in India is central to avoiding a catastrophic avoidable human tragedy like sub-Saharan Africa, where 15 million deaths have occurred since the beginning of the epidemic (Avert India, 2005). Likewise, successful crisis and risk communication efforts have to implicitly reflect and address several components of the HIV and AIDS discourses in India. A substantial reduction in the stigmatization of people living with HIV/AIDS is a key to any efficient and successful crisis and risk communication effort.

REFERENCES

AIDS Alert. (2002, September 17). *AIDS stigma forms an insidious barrier to prevention/care. HIV experts describe problem in India.* Sept, 17, 9, 111–113.

Avert India. (2005). *HIV and AIDS in India.* Retrieved from http://www.avert.org/aidsindia.htm

Bharat, S. (2001). *India: HIV and AIDS-related discrimination, stigmatization and denial. UNAIDS Best Practice Collection Key Material* (UNAIDS/01.46E). Geneva, UNAIDS. Retrieved November 20, 2004, from http://www.unaids.org/publications/documents/human/law/HR_India.pdf

Bharat, S., Aggeleton, P., & Tyrer, P. (2001). *India: HIV and AIDS-related discrimination, stigmatization, and denial.* Geneva: Joint United Nations Program on HIV AND AIDS. Retrieved from http://www.cbcihealth.com/HIV_india.pdf

Chatterjee, P. (2003). Spreading the word about HIV AND AIDS in India. *Lancet, 361*, 1526–1527.

Clinton, B. (2002). Speech at Global Business Council on HIV AND AIDS Awards dinner, New York.

Covello, V. T. (1992). Risk communication: An emerging area of health communication research. In S. A. Deetz (Ed.), *Communication Yearbook 15* (pp. 359–373). Newbury Park, CA: Sage.

Dutta-Bergman, M. J. (2004). The unheard voices of Santalis: Communicating about health from the margins of India. *Communication Theory, 14*(3), 237–263.

Ekstrand, M., Garbus, L., & Marseille, E. (2003). *HIV AND AIDS in India.* AIDS Research Institute: AIDS Policy Research Center, University of California, San Francisco. Retrieved from http://ari.ucsf.edu/ari/policy/countries.htm

Estrada, A., & Quintero, G. A. (1999). Redefining categories of risk and identity: The appropriation of AIDS prevention information and constructions of risk.

In W. N. Elwood (Ed.), *A handbook on AIDS, politics, and communication* (pp. 133–147). Mahwah, NJ: Erlbaum.

Fisbein, M. (2000). The role of theory in HIV prevention. *AIDS Care, 12*(3), 273–278.

Garret, L. (2000). *Lexicon of fear.* Retrieved June 16, 2005, from http://www.newsday.com/news/health/ny-aids20-africaaids2.story

Gates, B. (2002, November). Slowing the spread of AIDS in India. *The New York Times,* p. 19.

Gilmore, N., & Somerville, M. A. (1994). Stigmatization, scapegoating and discrimination in sexually transmitted diseases: Overcoming "them" and us. *Social Science and Medicine, 39*(9), 1339–1358.

Mandela, N. (2002, July). Speech delivered at the XIV International AIDS conference, Barcelona, Spain. Retrieved September 20, 2004, from http://www.kaisernetwork.org/aids2002/transcript/transcript_webcast_12_a2.html

National AIDS Control Organization. (2005). *HIV Estimates—2005.* Retrieved from http://www.nacoonline.org/facts_hivestimates05.htm

National Research Council. (1989). *Improving risk communication.* Washington, DC: National Academy Press.

Ogden, J., & Nyblade, L. (2005). *Common at its core: HIV-related stigma across contexts.* Washington, DC: International Center for Research on Women.

Page, S. (2002). *The metaphorical nature of HIV AND AIDS: An analysis of the coverage of HIV AND AIDS in four Zimbabwean Newspapers.* Paper presented at the Language, Literature and Discourse of HIV AND AIDS conference, Gaborone, Botswana.

Parrott, R. (2004). Emphasizing "communication" in health communication. *Journal of Communication, 54*(4), 751–787.

Pittam, J., & Gallois, C. (1997). Language strategies in the attribution of blame for HIV and AIDS. *Communication Monographs, 64,* 201–218.

Pittam, J., & Gallois, C. (2000). Malevolence, stigma, and social distance: Maximizing intergroup differences in HIV AND AIDS discourse. *Journal of Applied Communication Research, 28,* 24–43.

Raghavan, T. S. S. (2005, April). Now, not even a grave for AIDS-hit. *The Times of India.*

Rajalakshmi, T. R. (2004). India confronts AIDS. *Multinational Monitor, 25,* 12. Retrieved from http://multinationalmonitor.org/mm2004/122004/front.html

Rugamela, G. (2000). Fostering acceptance of HIV AND AIDS. Posted message. Retrieved on March 25, 2005, from http://www.edc.org.

Seeger, M. W., Sellnow, T. L., & Ulmer, R. R. (2003). *Communication and organizational crisis.* Westport, CT: Praeger.

Singhal, A. (2004). Combating HIV AND AIDS: Shout and shout and shout. *SPAN.*

Singhal, A., & Rogers, E. M. (2003). *Combating AIDS: Communication strategies in action.* Thousand Oaks, CA: Sage.

Singhal, A., & Vasanti, P. N. (2005). The role of popular narratives in stimulating the public discourse on HIV and AIDS: Bollywood's answer to Hollywood's Philadelphia. *South Asian Popular Culture, 3*(1), 3–15.

Solomon, S., Chakraborty, A., & Yepthomi, R. D. (2004). A review of the HIV epidemic in India. *AIDS Education and Prevention, 16*(Suppl. A), 155–169.

Solomon, S. et al. (2001). *The YRG CARE Model: An integrated continuum of prevention, and care and support for people living with HIV AND AIDS.* Washington DC: Horizons Program.

Sontag, S. (1991). *Illness as a metaphor. AIDS and its metaphors.* London: Penguin Books.

Thomas, M. (2005, February 16). In Satyagraha village, HIV+ woman is killed. *The Indian Express*, p. 1.

Turner, B. (1976). The organizational and interorganizational development of disasters. *Administration Science Quarterly, 21*, 378–397.

UNAIDS. (2002). *A conceptual framework and basis for action. FHI/UNAIDS Best practices in HIV/AIDS Prevention Collection.* Geneva: Author.

UNAIDS. (2004). *Report on the global AIDS epidemic: 4th global report.* Geneva: Author.

USAID. (2000). *USAID concept paper: Combating HIV AND AIDS stigma, discrimination and denial: What way forward.* Washington, DC: Author.

WHO. (2004). *AIDS watch: News from WHO South-East Asian region on STI, HIV and tuberculosis.* New Delhi: Author.

Witte, K. (1992). Putting the fear back into fear appeals. The extended parallel process model. *Communication Monographs, 59*, 329–349.

Wolf, A. (2002). *War, movement and food: The use of metaphors in discourses about AIDS.* Paper presented at conference on Language, Literature and Discourse of HIV AND AIDS, Gaborone, Botswana. Retrieved June 16, 2005, from http://bwenglish.tripod.com/conf.abstract.htm#kierkegaardina-Pointer-on—communication-for-the-Prevention-of-HIV-AIDS

You and AIDS. (2004). *The HIV AND AIDS portal for Asia Pacific: India at a glance.* Retrieved June 7, 2005, from http://www.youandaids.org/Asia%20Pacific%20at%20a%20Glance/India/index.asp#estimates

YRG Care. (2003). *YRG care base model report: Integrated prevention, care and support services.* Retrieved on December 8, 2005, from http://www.popcouncil.org/pdfs/horizons/yrgcrbsmdl.pdfs

PART V

Crisis Preparation and Response

11

APPLYING THE CRISIS AND EMERGENCY RISK COMMUNICATION (CERC) INTEGRATIVE MODEL TO BIOTERRORISM PREPAREDNESS

A Case Study

Deborah Ballard-Reisch   Kristen Clements-Nolle

Tia Jenkins   Teresa Sacks

Keri Pruitt   Kyle Leathers

This chapter applies the results of Nevada bioterrorism risk communication needs assessment initiatives to support and elaborate the CERC Integrative Model (Reynolds & Seeger, 2005). Results of Nevada initiatives strongly support Stages 1 to 3 of the model and elaborate and clarify constructs in all five stages. Implications of the applicability of the model to address a bioterrorism situation are discussed.

Communication is an essential component of any public health preparedness initiative and is rapidly becoming its own discipline in public health (Courtney, Cole, & Reynolds, 2003). Recognizing the importance of communication in the field of public health, Healthy People 2010 (2000) dedicated an entire chapter to its discussion. By definition, "health communication encompasses the study and use of communication strategies to inform and influence individual and community decisions that enhance health" (Dept. of Health and Human Services, 2000, p. 11-3). Crisis and risk communication are specific forms of public health communication that can be applied during any emergency or disaster situation. Crisis communication is designed to "prevent or lessen the negative outcomes of a crisis" (Coombs, 1999, p. 4); *risk communication* has been defined by Lundgren and McMakin (1998) as "the interactive process of exchange of information and opinions among individuals, groups, and institutions concerning a risk or potential risk to human health or the environment" (p. 329). Like other aspects of health communication, crisis and risk communication should take an audience-centered perspective to be effective. This means that the communicator must take into consideration an "audiences' preferred formats, channels, and contexts" (DHHS, 2000, p. 11-6) when developing communication strategies.

In October 2002, the Centers for Disease Control and Prevention (CDC) released an innovative online course based on a model of Crisis and Emergency Risk Communication (CERC) developed by Barbara Reynolds (2002). This course was designed as an introduction to "communication principles and tools as they relate to . . . emergency risk communication" to promote

> successful public, partner and stakeholder communication during an emergency situation. . . . Crisis and emergency risk communication is the attempt . . . to provide information that allows an individual, stakeholders, or an entire community to make the best possible decisions during a crisis . . . about their well being, and communicate those decisions, within nearly impossible time constraints, and ultimately, to accept the imperfect nature of choices as the situation evolves. (CDC, 2002, introduction)

Since that time, the CDC has developed a number of online and state-based trainings promoting the principle that effective communication is critical to effective crisis management.

In January/February 2005, Barbara Reynolds and Matthew Seeger published an application of the CDC CERC framework as an integrative model in the *Journal of Health Communication* for the purpose of emphasizing the communication tools necessary to manage natural disasters and public health emergencies throughout the developmental lifespan of a crisis. The purpose of this chapter is to apply the CERC framework, as elucidated by Reynolds and Seeger, to the unique case of bioterrorism preparedness. Findings from the University of Nevada, Reno, School of Public Health, public health preparedness needs assessment initiatives are used to empirically support, clarify, and modify the CERC model.

Specifically, the model's ability to characterize and address the information and response needs of diverse stakeholder groups before, during, and after a bioterrorist attack is evaluated and implications for public health preparedness are discussed.

UNIVERSITY OF NEVADA, RENO, SCHOOL OF PUBLIC HEALTH, BIOTERRORISM RISK COMMUNICATION NEEDS ASSESSMENT INITIATIVES

Gostin et al. (2002) defined *bioterrorism* as "the intentional use of a pathogen or biological product to cause harm to a human, animal, plant, or other living organism to influence the conduct of government or to intimidate or coerce a civilian population" (p. 623). The ultimate goal of bioterrorism is to produce fear and panic in the public, culminating in societal disruption or unrest. According to Henderson (1998), the potential for terrorists to use biological agents as weapons can produce results far more devastating than traditional forms of terrorism (e.g., bombings, shootings, hijackings, assassinations, release of chemical agents).

Bioterrorism is so potentially devastating because, unlike a traditional terrorist attack, in which a specific event is observed to have taken place (e.g., planes flying into the World Trade Towers on September 11, 2001), bioterrorism is clandestine. A biological agent could be released that would take weeks to detect. By then, those immediately impacted could have spread contagion widely, compounding the severity of the situation and potentially creating an epidemic (see Hamburg, 2000). Identifying a bioterrorism attack is much like putting together a jigsaw puzzle—the picture emerges as more pieces are added.

Recognizing the need to improve the public health infrastructure to respond to the unique aspects of bioterrorism, the CDC developed cooperative agreements with state and local health departments to address planning, surveillance, and epidemiology; laboratory capacity; communications and information technology; risk communication and public information dissemination; and education and training (CDC, 2005). To determine the risk communication and public information needs of various stakeholder groups throughout Nevada, the State Health Division contracted with the University of Nevada, Reno, School of Public Health to conduct a comprehensive needs assessment. From October 16, 2002, through October 15, 2004, three methodologies were used to gain information regarding risk communication needs in the event of a bioterrorism situation: focus groups, key informant interviews, and an online survey of Nevada's public school teachers and administrators. Stakeholders were recruited from northern

Nevada (primarily the Reno/Sparks area and Carson City), southern Nevada (primarily the Las Vegas area), and the state's 15 rural counties; 688 participants took part in initiative activities. This research received human subjects exemptions from the University of Nevada Reno, Social and Behavioral Institutional Review Board.

Focus Groups

Twenty-two needs assessment focus groups were conducted with a diverse range of stakeholders, including emergency responders (three groups), emergency medical providers (one group), elected and appointed public officials (three groups), tribal health representatives (two groups), casino safety officers or risk managers (two groups), and diverse segments of the public population such as homeless adults, college students, rural adults, urban adults, seniors, African Americans, and Hispanics/Latinos (11 groups); 195 people participated in focus groups. (For a more complete discussion of participant demographics, topics areas covered, and data analysis, see Clements-Nolle, Ballard-Reisch, Jenkins, & Todd, 2005).

Key Informant Interviews

Key informant interviews were conducted with 23 media representatives from across the state, 17 county health officers, and 42 community health nurses representing each county in Nevada. Nine to 11 key informants from each of six co-cultural groups identified by the Nevada State Health Division were interviewed (African Americans, Asians and Pacific Islanders, Hispanics/Latinos, Native Americans, persons with disabilities, and seniors), as were 91 trainees of a statewide Critical Incident Stress Management program; 233 key informants were interviewed. (For a more complete discussion of participant demographics, topics areas covered, interview protocols, and data analysis, see Clements-Nolle et al., 2005; Jenkins, Ballard-Reisch, Clements-Nolle, & Caloiaro, 2005).

Online Survey of Nevada Public School Teachers and Administrators

A Web-based, online survey was completed by 127 public school teachers and 133 school administrators from throughout the state of Nevada; 260 participants took part in the online survey. (A more complete discussion of participant demographics, survey development, survey content, and data analysis can be found in Pruitt, Ballard-Reisch, Clements-Nolle, & Letner, 2004).

OVERVIEW OF THE CRISIS AND EMERGENCY RISK COMMUNICATION (CERC) MODEL

Integrating health, risk, crisis, and disaster communication research, the CERC model is composed of five stages that articulate communication priorities within a developmental approach to emergency crisis management (Reynolds & Seeger, 2005). The model is advanced as a tool for health communicators to assist in managing complex public health threats, including crises and disaster events as well as emergent threats to public health such as infectious disease outbreaks and potential bioterrorist attacks.

The five stages are: (1) precrisis, which emphasizes risk messages, warnings, and preparations, including communication and education programs to facilitate risk monitoring. The precrisis stage works toward enhancing the public's knowledge and understanding of risk, developing interagency and interorganizational collaborations, and testing messages for use in the event of an emergency/crisis; (2) initial event, which involves rapid communication to the public and stakeholder groups to reduce uncertainty, clarify what has occurred, facilitate response, and build public self-efficacy and response capacity; (3) maintenance, which is concerned with the ongoing facilitation of the situation, including continued uncertainty reduction to enhance decision making; (4) resolution, which deals with communicating with the public regarding ongoing clean-up, remediation, recovery, and rebuilding efforts, and addresses issues of cause, blame, responsibility, and adequacy of response; and (5) evaluation, which emphasizes lessons learned and new understandings of risk, and procedural modifications to improve future response, crisis communication, and crisis response capability (see Table 11.1 for graphic representation of Reynolds & Seeger's [2005] working model of CERC).

TABLE 11.1.
A Working Model of CERC (Reynolds & Seeger, 2005)

I. Precrisis (Risk Messages, Warnings, Preparations)

Communication and education campaigns targeted to both the public and the response community to facilitate:

- Monitoring and recognition of emerging risks
- General public understanding of risk
- Public preparation for the possibility of an adverse event
- Changes in behavior to reduce the likelihood of harm (self-efficacy)

TABLE 11.1.
A Working Model of CERC (Reynolds & Seeger, 2005) *(continued)*

- Specific warning messages regarding some eminent threat
- Alliances and cooperation with agencies, organizations, and groups
- Development of consensual recommendations by experts and first responders
- Message development and testing for subsequent stages

II. Initial Event (Uncertainty Reduction, Self-Efficacy, Reassurance)

Rapid communication to the general public and to affected groups seeking to establish:

- Empathy, reassurance, and reduction in emotional turmoil
- Designated crisis/agency spokespersons and formal channels and methods of communication
- General and broad-based understanding of the crisis circumstances, consequences, and anticipated outcomes based on available information
- Reduction of crisis-related uncertainty
- Specific understanding of emergency management and medical community responses
- Understanding of self-efficacy and personal response activities (how/where to get more information)

III. Maintenance (Ongoing Uncertainty Reduction, Self-Efficacy, Reassurance)

Communication to the general public and to affected groups seeking to facilitate:

- More accurate public understandings of ongoing risks
- Understanding of background factors and issues
- Broad-based support and cooperation with response and recovery efforts
- Feedback from affected publics and correction of any misunderstandings/rumors
- Ongoing explanation and reiteration of self-efficacy and personal response activities (how/where to get more information) begun in Stage II.
- Informed decision making by the public based on understanding of risks/benefits

IV. Resolution (Updates Regarding Resolution, Discussions about Cause and New Risks/New Understandings of Risk)

Public communication and campaigns directed toward the general public and affected groups seeking to:

- Inform and persuade about ongoing clean-up, remediation, recovery, and rebuilding efforts
- Facilitate broad-based, honest, and open discussion and resolution of issues regarding cause, blame, responsibility, and adequacy of response

- Improve/create public understanding of new risks and new understandings of risk as well as new risk avoidance behaviors and response procedures
- Promote the activities and capabilities of agencies and organizations to reinforce positive corporate identity and image

V. Evaluation (Discussions of Adequacy of Response; Consensus About Lessons and New Understandings of Risks)

Communication directed toward agencies and the response community to:

- Evaluate and assess responses, including communication effectiveness
- Document, formalize, and communicate lessons learned
- Determine specific actions to improve crisis communication and crisis response capability
- Create linkages to precrisis activities (Stage I)

APPLICATION OF THE CERC MODEL TO A BIOTERRORISM SITUATION: NEEDS ASSESSMENT RESULTS

Results from the bioterrorism risk communication needs assessment conducted by the University of Nevada, Reno, School of Public Health support, clarify, or enhance all five stages of the CERC model. Each stage of the model and supporting data from the Nevada initiatives are discussed in turn.

Precrisis (Risk Messages, Warnings, Preparations)

This stage, typically viewed as risk communication, involves planning efforts and most development work. During the precrisis stage, Reynolds and Seeger (2005) noted the need for: (1) monitoring and recognition of potential risks, including the preparation of the public to both understand risk and prepare for a possible adverse event; (2) developing alliances and cooperation among agencies, organizations, and groups; and (3) developing and testing educational and informational materials, including specific warning messages geared toward promoting individual self-efficacy. Our results also highlight the importance of (4) pre-event education and training, and (5) the development and testing of emergency response plans.

Monitoring and Recognition of Potential Risks. Being able to detect emerging risks is critical in the event of a bioterrorism attack. Because bioterrorism is emergent, it is not anticipated that there will be a clearly identifiable event to mark its initiation. Results of our needs assessment investigations indicate that the ability of frontline health providers, including public health professionals, medical professionals, and emergency first responders, to recognize and diagnose is extremely important. Medical providers, first responders, community health officers, and nurses wanted to play a role in early detection and reporting activities and were willing to receive the training necessary to do so. They also noted a need for accurate surveillance and communication systems as well as updated laboratory equipment so that rapid identification and response are possible. The public wanted to know what to look for and to whom they should report any suspicious incidents or troubling symptoms.

Alliances and Cooperation With Agencies, Organizations, and Groups. Effective response to a bioterrorism situation requires the coordinated action of diverse agencies, organizations, and groups to create a unified approach to risk assessment and response. Our research demonstrated significant concern among public and professional stakeholders that preparedness planning to date has been occurring in isolation and, if an event were to occur, key stakeholders would be acting independently. The development of alliances and partnerships and an integrated response strategy were identified by participants as critical in pre-event planning for five reasons: (1) They allow for the development of a coordinated, responsive public health infrastructure; (2) a coordinated infrastructure is critical for coordinated response; (3) coordination allows for the clarification of roles and responsibilities; (4) coordination ensures that, in the event of a bioterrorism situation, experts will speak with one voice; and (5) many groups already have effective information-dissemination mechanisms that could be utilized in case of a bioterrorism or other public health emergency.

Participants stressed that pre-event collaborations and relationships with the media must also be developed to promote trust and cooperation between the media and other stakeholder groups. Collaborative efforts with the media should promote: (1) open, timely communication to the media; (2) media access to official spokespersons; (3) establishment of pre-event relationships among public health agencies, official spokespersons, and the media based on trust and cooperation; and (4) collaboratively developed information-dissemination protocols and formats.

Pre-event alliances and cooperation must also involve first responders and medical providers, including physicians and community health nurses who are responsible for incident identification, diagnosis, surveillance, and treatment. Additionally, one largely overlooked group in preparedness efforts to this point has been mental health professionals.

Because of their public and policy roles, public officials will be involved in preparedness and response activities. In collaboration with public health officials, elected and appointed officials noted that they will be responsible for determining the interface among individual and collective rights and protections, as well as resource allocation and policy implementation and enforcement. To fulfill their roles, they need: (1) pre-event communication regarding the needs of specific stakeholder groups in terms of policy and resource allocation; (2) consistent involvement in planning activities; (3) a clear indication of the mechanics of how and through whom they will be informed should an event emerge and throughout the crisis management process; and (4) appropriate training in threat assessment, crisis management, and rebuilding postcrisis infrastructures.

Many community groups, including churches, schools, cultural organizations, senior centers, the Division of Aging Services, the Office of Veterans Affairs, and disability resource centers, have access to and understand unique segments of the population. In addition, they have their own infrastructure in place to get information to their constituents and communities. Public participants suggested that, in a crisis, these networks would be highly beneficial in disseminating information in a timely manner using appropriate language and format.

Perhaps the most universal finding throughout Nevada's needs assessment initiatives was participants' belief that the public must be viewed as partners in preparedness and response. Participants were explicit that the paternalistic view that authorities should withhold information to avoid panic was a misguided and potentially dangerous strategy that would decrease trust and increase the likelihood of panic. The public cannot easily be contained in a crisis and would be a valuable partner in the mobilization of response. This would be particularly true if they: (1) adequately prepared for a crisis event so they knew what to do, where to go, what to report, and so on; (2) received accurate, timely, ongoing information regarding the situation as it emerges; and (3) are empowered to mobilize extant resources to support response. School administrators, for example, were well aware of the need to actively include parents and students in planning and preparedness efforts and emergency plan drills. Examples of extant resources on which the public could draw to aid response include the use of church phone trees; parent–teacher organizations at local schools to communicate information and coordinate response initiatives; and meals-on-wheels programs to identify, reach, and get needed information and supplies to frail or homebound seniors or individuals with disabilities.

Developing and Testing Educational and Information Materials and Dissemination Strategies. Reynolds and Seeger (2005) identified the need for message development and testing for pre-event education and subsequent stages of crisis management. Public participants in Nevada's needs assessment initiatives were interested in tailored, personally relevant, and accurate educational information regarding: (1) elements of potential situations (including biological agents, transmission, signs and symptoms, and prevention measures); (2) by whom and how the situation will be managed; (3) how and through what channels information will be communicated to the public; and (4) where they can go for more information. Equally important as the need for message development is the need to determine appropriate channels and methods of communication, ensure that messages are targeted to particular audiences, and designate official spokespersons.

An overwhelming percentage of participants agreed that information must be released to the public and to professional stakeholders in a timely manner, using predetermined formats, through previously identified dissemination strategies. Participants also recommended pilot testing of information type (what questions/issues would be addressed), message format, and the channels through which the information is disseminated.

Respondents were clear that messages that do not reach their target audiences, or fail to do so in a timely and efficient manner, hurt preparedness and response efforts. In developing appropriate messages, planners need to consider: (1) the capacity of diverse target audiences in message preparation, (2) standardization of information formats, (3) preparation of multilingual materials, and (4) the development of both high- and low-technology information-dissemination strategies. Participants recommended that planners anticipate questions, formulate responses, and leave blanks for additional information to be filled in depending on the characteristics of a particular situation.

In addition, needs assessment respondents believed that the designation of emergency spokespersons who were well known and easily accessible throughout the precrisis stage would enhance credibility and confidence in those spokespersons in the event of an actual bioterrorism emergency. Spokespersons should be: (1) knowledgeable; (2) able to establish proactive relationships with the media and diverse stakeholder groups; (3) accessible to the media and diverse stakeholder groups; (4) open and direct communicators; and (5) appropriate for diverse populations. As representatives from some co-cultural populations such as Native Americans and Hispanics indicated, different groups have different cultural and communication norms and expectations and may lack trust in government and authority figures. Therefore, it is necessary to enlist culturally competent spokespersons who populations will trust and with whom they can identify.

Pre-event Education and Training. A wide variety of public and professional groups including first responders, medical providers, media representatives, public officials, and the general public in our assessment, highlighted the need for pre-event education and training in order to be prepared for a potential bioterrorism situation.

Both public and professional participants indicated a desire to know what to do to protect themselves and those they love, or for whom they are responsible, during a bioterrorism situation. Participants thought that the more prepared they felt in advance, and the clearer the procedures in place to direct them in response, the more likely they would respond effectively and feel a sense of control during an actual bioterrorism situation. Participants universally expressed the opinion that precrisis education would ultimately lead to less fear and panic if a bioterrorism event were to occur.

Regardless of the needs assessment methodology used, all stakeholders indicated that their first access to information would be through the media. They believed that for the media to effectively educate and communicate with communities, they had to be adequately prepared, and that this was best done before a situation emerges. Participants stressed the need for ongoing pre-event education of the media. Areas of media training included: (1) infectious disease, (2) bioterrorism, (3) risk communication, (4) event management, and (5) the role the media can play in information dissemination.

In addition to the public and the media, professional stakeholders such as emergency first responders, medical providers and community health nurses, and casino safety officers identified a variety of needs for pre-event education and training programs in order to assist them in preparing for and understanding responsibilities and the new roles they would be required to play in the event of a bioterrorism situation. Stakeholder groups who would likely be called to emerging public health threats believed that such training would reduce their uncertainty about and unfamiliarity with bioterrorism and would clarify the roles they would play if a bioterrorism event occurred. Professional participants identified needs for training in: (1) recognition, identification, and reporting; (2) isolation and containment; (3) public education and risk communication; and (4) stabilization and recovery.

Development and Testing of Emergency Response Plans. In addition to collaboration with specific groups and focused training and education programs, participants articulated the need to coordinate and rehearse response plans and standard operating procedures. Concern over a lack of coordination among response plans and standard operating procedures was significant for both public and professional respondents who saw collaboration and integration of plans as critical to effective response. Professional participants such as casino safety officers and public officials requested an emergency plan template and leadership from the state and county health departments in coordinating the efforts of diverse groups that would play a role in response. School administrators also expressed a desire for a template so that all schools in the state would have comparable plans that adequately addressed preparedness and response issues. Numerous participants also noted that, no matter how good a plan appears on paper, the proof of its effectiveness is in the testing. They recommended regular testing of response plans and standard operating procedures through mock drills and scenarios that would involve stakeholders who historically have not been included, such as the media and the public.

INITIAL EVENT (UNCERTAINTY REDUCTION; SELF-EFFICACY; REASSURANCE)

During an initial event, risk and crisis communication should aim to reassure the public, reduce uncertainty, and promote self-efficacy (Reynolds & Seeger, 2005). Our results demonstrate that such outcomes can most effectively be achieved by: (1) effective response mobilization, and (2) open, timely, and accurate information dissemination.

Response Mobilization. Collaborative emergency plans and standard operating procedures for first responders need to be implemented so that coordination among agencies and organizations can occur to effectively analyze the situation and initiate response. If recognition, response, and risk communication plans are developed, evaluated, and updated before an event occurs, response mobilization should be comprehensive and efficient. One potential problem during this stage of a crisis is dealing with the "worried well" who have the potential to overwhelm medical and response capacity. Mental health providers could play a role both in managing the fears of community members as well as triaging those with physical symptoms, those with reasonable concerns, and those experiencing psychological distress as a result of the event.

Information dissemination. During a bioterrorism situation, the information-dissemination strategies developed during the precrisis stage would be tested. When discussing information dissemination at the onset and throughout a crisis, Nevada participants overwhelmingly reinforced a need for: (1) open and honest communication, (2) consistent information across sources (the need for governmental officials to communicate with a unified voice), (3) avoidance of hedging on the part of officials, (4) frequent updates as the situation progresses, (5) utilization of both low- and high-tech information-dissemination approaches, (6) targeted and appropriate messages directed toward specific audiences, and (7) utilization of community groups and agencies to disseminate information.

Participants consistently expressed their desire for open and honest communication during a public health emergency. As stated earlier, it was generally thought that withholding information erodes public trust, and this could greatly hamper the management of a crisis situation. Participants also stated that public health and government officials need to communicate in one voice and, where possible, use the same materials so that consistent, coherent messages are communicated. If official sources communicate contradictory or different information, public confidence in the knowledge and competence of experts is damaged, as is the public's belief in the ability of the public health infrastructure to manage the crisis. In addition, it is critical that spokespersons communicate openly and that they are not perceived as evasive. According to our participants, responses like "I'm not at liberty to respond to that question" or "I don't have that information" damage trust. Spokespersons should state what they know, what efforts are being made to find out additional information, and when additional information is expected to be available.

Public and professional stakeholders also expressed the desire for frequent updates as information becomes available using previously identified spokespersons, and channels and methods of communication, including both high- and low-technology dissemination methods. Suggested dissemination modalities included: (1) a 1-800 number staffed 24-7 by knowledgeable public health professionals who can answer and/or triage calls to appropriate professionals; (2) a well publicized, easily accessible State bioterrorism Web site; (3) fliers, newsletters, and informational brochures for technologically limited or isolated communities; (4) interpersonal and community-based information dissemination through churches, schools, and other community and cultural organizations; (5) community meetings or door-to-door information dissemination; (6) closed captions, large print, audio, TTD/TTY for persons with disabilities and seniors; and (7) multilingual media and material resources.

Participants repeatedly reinforced the need to tailor both messages and dissemination strategies to the needs of specific populations. For example, there is a large population of first language Spanish speakers in Nevada. Participants indicated that in times of stress they tend to revert to their first language. Materials in Spanish would assist the population in both comprehension and appropriate response. Similarly, seniors (Nevada has the fastest growing senior population in the United States; U.S. Census Bureau, 2000) and individuals with disabilities may have sensory limitations that necessitate that messages be delivered using larger font, slower ticker tapes on newscasts or closed captioning, TTD/TTY services, Internet communication, and so on in order to be effective. Nevada's rural communities and Native American tribes prefer print (flyers, brochures, posters, fact sheets) and face-to-face dissemination strategies due to the lack of availability and reliability of TV and Internet service and cultural traditions.

Maintenance (Ongoing Uncertainty Reduction, Self-Efficacy, Reassurance)

The groundwork for appropriate ongoing crisis management was ideally set in the precrisis and initial event stages. However, quality communication remains critical to the ongoing management of an emergency situation. This is particularly true for bioterrorism and other public health emergencies that can be expected to continue for a period of time. To be effective, continued communication should: (1) clarify the background factors and contextual issues surrounding the context, (2) allow for the management of misunderstandings and rumors, (3) promote broad-based support and cooperation with response activities and recovery efforts, (4) enhance public understanding of ongoing risks, (5) keep the public informed about where to go for additional information, and (6) promote decision making based on understanding of risks and benefits (Reynolds & Seeger, 2005).

In addition, during the maintenance phase of a crisis situation, the public's interest in volunteerism and community assistance may increase. Nevada respondents indicated that once personal health and safety issues were managed, they wanted to know what they could do to assist response, recovery, and rebuilding efforts. Public groups were particularly interested in where they could go to volunteer their time and how they could help others. Mental health providers noted the importance of giving people something to do so they can feel useful and productive because this is the point in crisis management when stress, depression, and posttraumatic stress disorder (PTSD) symptoms might emerge.

Resolution (Updates Regarding Resolution, Discussions About Cause and New Risks/New Understandings of Risk)

Although Nevada needs assessment initiatives did not directly deal with this stage, it is clear that quality communication strategies and tactics identified in Stages 1 to 3 will continue to enhance response and recovery efforts and are crucial to the following goals of resolution: (1) informing and persuading stakeholders regarding clean-up, remediation, recovery, and rebuilding efforts; (2) facilitating discussion and resolution of issues regarding cause, blame, responsibility, and adequacy of response; and (3) enhancing public understanding of risks, risk avoidance behaviors, and response procedures (Reynolds & Seeger, 2005). Additionally, at this stage, Nevada respondents indicated that it would be important to acknowledge the collaborative roles and contributions of agencies, organizations, the public, and other stakeholders in response efforts.

Evaluation (Discussion of Adequacy of Response, Consensus About Lessons and New Understandings of Risks)

Reynolds and Seeger (2005) direct this final stage of the CERC model to agencies and the response community, which they task with: (1) evaluating the effectiveness of response including communication effectiveness; (2) documenting, formalizing, and communicating lessons learned; (3) determining specific actions to improve crisis communication and crisis response capability; and (4) creating linkages to precrisis activities (Stage 1).

Professional respondents in the Nevada needs assessment activities also noted the need for: (1) consistently updated standard operating procedures and emergency response plans for first responders, mental health responders, and other agencies and organizations that will take part in future incident response; (2) the maintenance of continually updated and revised contact lists and phone trees; (3) ongoing training to keep collaborations intact and roles and responsibilities clear and well articulated; (4) ongoing mock scenarios and drills to keep skills sharp; and (5) ongoing evaluations of risk communication and response plans.

Finally, public perceptions of the effectiveness of response efforts are critical as well. As partners throughout the process of crisis management, their assessment of response activities as well as recommendations for future capacity building should provide valuable insights, and their views should be solicited during this stage.

CONCLUSION

This chapter applied the CERC model to a bioterrorism situation using data from Nevada public health preparedness needs assessment activities. While providing support for the five stages in the Reynolds and Seeger (2005) model, analysis also allowed for the elaboration of the model, particularly in Stages 1 to 3 which are essential for bioterrorism planning and response. As our results demonstrate, precrisis Stage 1 initiatives are critical. If done well, Stage 1 initiatives lay the foundation for all other stages in the CERC model. If precrisis initiatives are not properly developed, managing a crisis once a bioterrorism event has emerged is much more difficult and response effectiveness significantly hindered. When preparing for a possible bioterrorism situation, it is critical that the precrisis stage (Stage 1) focuses on pre-event training and education for the public and professional stakeholders, including the development of pre-event collaborations and alliances, which are essential to define roles and responsibilities, coordinate planning efforts, and avoid duplication. Collaborations should involve diverse stakeholder groups such as the media, public officials, health providers, and, most important, the public. Well before an event occurs, quality emergency response plans and standard operating procedures must be developed and evaluated. In addition to the development of bioterrorism education and information materials, dissemination strategies must be developed and pilot tested. Our results also highlight the need to identify official, credible, and knowledgeable spokespersons during the precrisis stage, rather than the initial event stage to ensure that the public knows and trusts the spokespersons *before* an event actually occurs.

During response to an initial crisis event (Stage 2), response mobilization and the communication of accurate, timely, and complete information is critical. During this period, hedging or withholding information to avoid panic will decrease trust in official spokespersons and those responsible for response and may ultimately hinder response efforts. There is also a need for consistent messages across sources and mediums of dissemination, timely updating of information through appropriate channels, and the utilization of community groups and agencies for information dissemination.

During the maintenance stage of a crisis or emergency situation (Stage 3), there is a need to sustain efforts from prior stages and clear up any confusion or misunderstandings. Involving the public as partners in all phases of crisis and emergency risk communication is important, but our results suggest that the maintenance stage may be a particularly important time to promote public volunteerism efforts.

Although our needs assessment data did not provide much information that allowed for the application or refinement of the resolution stage (Stage 4), our findings do support the need for ongoing evaluation (Stage 5). Initiatives geared toward evaluation should include the consistent updating of emergency response plans and standard operating procedures, the maintenance of updated and revised contact lists and phone trees, ongoing mock scenarios and drills, evaluation of risk communication and response plans, and assessment of response effectiveness.

Effective communication before, during, and after public health emergencies is the foundation for appropriate response. The CERC model provides an integrative framework for mobilizing effective risk and crisis communication in situations that threaten the public's health. Continued refinement and application of the CERC model to diverse public health crises and emergencies will improve its ability to promote effective preparedness and response.

ACKNOWLEDGMENTS

Funding for this project was provided by the Nevada State Health Division through CDC Public Health Preparedness Grants U90/CCU916964-03-01 and U90/CCU916964-04-01. In addition, the authors would like to thank Martha Framsted, Public Information Officer, Nevada State Health Division; and other members of our student research team: Jessey Bargmann, Carla Bellusa, Joyce Letner, Bryce Putnam, Dr. Timothy Ricks, Marcia Sarratea, Anna Treacy, and Jennifer Vinson.

REFERENCES

Centers for Disease Control and Prevention (October, 2002). *Crisis and emergency risk communication.* Atlanta, GA: Centers for Disease Control and Prevention. Retrieved July 20, 2004, from www.cdc.gov.

Centers for Disease Control and Prevention (2005, January 13). *Continuation guidance for cooperative agreement on public health preparedness and response for bioterrorism—budget year five.* Available at http://www.bt.cdc.gov/planning/continuationguidance/index.asp

Clements-Nolle, K., Ballard-Reisch, D. S., Jenkins, T. & Todd, R. L. (2005). Nevada's academic-practice collaboration: Public health preparedness possibilities outside an academic center. *Public Health Reports, 120*(Supplement 1), 100–108.

Coombs, W. T. (1999). *Ongoing crisis communication: Planning, managing, and responding.* Thousand Oaks, CA: Sage.

Courtney, J., Cole, G., & Reynolds, B. (2003). How the CDC is meeting the training demands of emergency risk communication. *Journal of Health Communication, 8*, 15–16.

Department of Health and Human Services. (2000). *Healthy people 2010* (2nd ed.). Washington, DC: Author.

Gostin, L. O., Sapsin, J. W., Teret, S. P., Burris, S., Mair, J. S., Hodge, J. G., & Vernick, J. S. (2002). The model state emergency health powers act: Planning for and response to bioterrorism and naturally occurring infectious diseases. *JAMA, 288*, 622–628.

Hamburg, M. A. (2000). Bioterrorism: A challenge to public health and medicine. *Journal of Public Health Management Practice, 6*(4), 38–44.

Henderson, D. A. (1998). Bioterrorism as a public health threat. *Emerging Infectious Diseases, 4*(3), 488–492. Retrieved August 29, 2003, from http://www.cdc.gov/ncidod/EID/vol14no3/henderson.htm.

Jenkins, T., Ballard-Reisch, D., Clements-Nolle, K., & Caloiaro, D. (2005, November 5–9). *Pragmatic training solutions in mental health disaster response: Findings from Nevada's critical incident stress management training assessment.* Paper presented at the 133rd annual meeting of the American Public Health Association, New Orleans, LA.

Lundgren, R., & McMakin, A. (1998). *Risk communication: A handbook for communicating environmental, safety, and health risks.* Columbus, OH: Battelle.

Pruitt, K., Ballard-Reisch, D., Clements-Nolle, K., & Letner, J. (2004, November 6–10). *Bioterrorism on a school day: Identifying the emergency preparedness needs of Nevada's public schools.* Poster presented at the 132nd annual meeting of the American Public Health Association, Washington, DC.

Reynolds, B., & Seeger, M. (2005). Crisis and emergency risk communication as an integrative model. *Journal of Health Communication, 10*, 43–55.

U.S. Census Bureau. (2000). *Profile of General Demographic Characteristics for Nevada.* Table DP-1.

12

PARALLEL PROCESS MODEL AND GOVERNMENT PREPAREDNESS MESSAGES

Beyond Duct Tape and Plastics Sheeting

W. Timothy Coombs

At the end of 2005, the Department of Homeland Security (DHS) launched a new phase in its preparedness campaign. A new direction was needed because the previous campaign failed. People were not preparing for emergencies nor would they be likely to listen to government recommendations during a biological or dirty bomb attack. This chapter places the failure within the context of the Extended Parallel Process Model (EEPM). The government and the media are effective at instilling fear, a motivating factor, but not in developing feelings of efficacy. People do not seem to appreciate that government plans can work and/or that they can take actions that will help to protect themselves. The preparedness messages need to emphasize efficacy content if more Americans are to take preparedness actions and be willing to follow government recommendations during an emergency.

In October 2004, Vice President Dick Cheney told the American people, "The biggest threat we face now as a nation is the possibility of terrorists ending up in the middle of one of our cities with deadlier weapons than have ever before been used against us—biological agents or a nuclear weapon or a chemical weapon of some kind—to be able to threaten the lives of hundreds of thousands of Americans. . . . You have to get your minds around that concept" (Archibold, 2004, p. 20). Cheney's October alert is part of a long list of messages warning Americans of another possible terrorist attack. Through the Ready.gov Web site and DHS's public service campaign, the federal government has been trying to spread the word about preparedness for terrorist attacks—what can be called the *terrorism preparedness campaign*.

Unfortunately, the preparedness messages have failed to get traction among the America people. In 2003, a terrorist warning sent East Coast residences on a duct tape and plastic sheet buying spree. Apparently some of the people were aware of using the products to "shelter-in-place." But no message to shelter-in-place had been given, only a vague message about a potential attack. More disturbing was the September 2004 W. K. Kellogg Foundation-sponsored study entitled, *Redefining Readiness*. The study examined how Americans might react to a smallpox outbreak or a dirty bomb. Only 43% of the American people would go to a smallpox vaccination site if an outbreak was reported and government officials told them to get vaccinated. A full 55% would wait for more information and then decide. If people were told a dirty bomb had been detonated and they should stay inside where they were, only 68% would comply. Another 11% would leave the building while 20% would wait for more information (Redefining, 2004). The conclusion: A sizable segment of the U.S. population would disregard government protective instructions during a terrorist incident.

There is empirical evidence that the terrorism preparedness campaign is ineffective. The question we must explore is "Why?" If we can determine why the terrorism preparedness campaign is failing, we may be able to develop a more effective campaign. In turn, a more effective campaign will ensure greater survival rates if or when another terrorist attack on U.S. soil occurs. The EEPM is an effective tool for developing and assessing risk communication messages (Witte, Meyer, & Martell, 2001). This chapter posits that EEPM is a useful framework for diagnosing and improving terrorism preparedness messages. The analysis unfolds in five sections. The first section highlights key insights from the Kellogg Foundation's *Redefining Readiness* report. The second section outlines the EEPM model. The third and fourth sections examine the use of fear and efficacy in the government's terrorism preparedness campaign. The fifth section discusses implications from the analysis.

CONCERNS FROM REDEFINING READINESS

The Kellogg Foundation's *Redefining Readiness* report was a shock to the emergency preparedness community. It was a featured item in one of the community's major Internet-based newsletters. The initial shock is that a large segment of the U.S. population would not follow the government's plans for a smallpox or dirty bomb terrorist incident. People would disregard government messages about how to respond to these two terrorist scenarios. The results are more disturbing when moving beyond the headlines to the study specifics. There is an amazing lack of awareness of the government's terrorism preparedness efforts, which may help explain the lack of confidence in the government's attempts to combat a terrorism incident.

When asked whether they were aware of any government plans developed for either small pox or a dirty bomb, 74% of the respondents reported they were unaware. Of the 26% who said they were aware, 44% said they knew little and 13% said they knew nothing when asked how much they knew about the way government agencies would handle these two terrorism scenarios. On the community level, 32% thought their community was a little prepared for a terrorist attack and 23% thought the community was not prepared at all (Redefining, 2004). This is a stunning lack of awareness about the government's terrorism preparedness on both the federal and community levels. It is even more surprising in that DHS is devoting considerable time and resources to emergency preparedness, including a public service campaign to increase awareness. The remainder of this chapter explores why these results occurred and how we might turn these numbers around.

THE EXTENDED PARALLEL PROCESS MODEL

The government and other organizations have tried to improve the health of the U.S. population around a variety of topics ranging from seat belt usage to colon cancer. These health messages are delivered though a number of channels including public service announcements (PSAs). Terrorism preparedness qualifies as a health message. The objective of the campaign is to reduce the risks associated with a terrorist attack. Most of these risks are health related. Knowing what to do should reduce the loss of human life and the number of injuries. Researchers have long sought to understand how to design effective health messages. An effective health message is one that motivates the listener to make the desired changes in her or his behavior. For example, a person wears a seat belt or has a blood pressure (BP) check. The EPPM provides a theoretical framework for designing effective health messages (McMahan, Witte, & Meyer, 1998; Witte et al., 2001). EPPM then serves as the analytic tool to help evaluate the failure of the terrorism preparedness campaign. This section examines the development and structure of the EPPM

EPPM Background

EPPM is rooted in fear appeals—messages that warn of negative consequences if certain behaviors are not enacted. There has been over 45 years of research and application of fear appeals to health messages. A fear appeal includes a threat, something bad will happen, and a recommended response—what you can do to prevent that something bad. For example, you are more likely to die in an automobile accident if you are not wearing a seat belt (threat), so wear a seat belt (recommended response). The idea is that the emotions created by fear motivate people to take action. EPPM draws on a variety of fear appeal research with a strong connection to Leventhal's (1970) parallel process model (Witte et al., 2001). Leventhal posited that emotional responses to fear prompt a cognitive appraisal. People think about the situation and decide to use either a fear control process or a danger control process. In the fear control process, people take actions to control the fear, including efforts to avoid thinking about the fear. The emphasis is on controlling the fear. In the danger control process, people take action designed to alleviate the threat. The emphasis is on enacting behavioral alterations designed to protect themselves from the threat (Witte et al., 2001). Ideally, health messages should promote danger control. The messages should lead people to enact behaviors that will protect them from the threat.

The EPPM outlines how people cognitively appraise threats and the effects of these threats on their behavioral choices. A health message about a risk will initiate two cognitive appraisals: threat appraisal and efficacy appraisal. In turn, the results of these appraisals determine how the person will respond to the health message. A threat is appraised for perceived susceptibility and perceived severity. Perceived susceptibility assesses whether the threat is relevant to the individual: Can this risk really affect them? Perceived severity is the amount of danger presented by the threat—whether it can inflict great harm. If people believe they are not vulnerable to a threat and/or it can inflict only minor harm, they will not react to the health message. However, if the threat is viewed as serious and/or the person feels vulnerable, she or he is motivated to act. The more intense the perceived threat, the more motivated the individual is to act (McMahan et al., 1998; Witte et al., 2001).

Once motivated, people then assess the efficacy of the recommended response—what they are told to do to protect them from harm. The evaluation of efficacy involves self-efficacy and response efficacy. Self-efficacy is whether people feel they can enact the recommended response. Perceived self-efficacy is high when people believe they can perform the recommended response. Response efficacy is the perception that the recommended response will actually function to avoid the threat. People determine whether they think the recommended response will be effective. When self-efficacy and response efficacy are perceived to be high, people take the recommended action to control the threat. This is the protection motivation response. When perceptions of self-efficacy and/or response efficacy are low, people feel they cannot control the risk. People will try to control their fear through denial or avoidance. This is the defensive motivation response (Witte et al., 2001). The protection motivation is the desired response because people are enacting the appropriate behaviors to reduce the threat. To be effective, a health message must have a threat that is perceived as severe and relevant to the audience while offering a course of action that people believe they can perform and that they believe will avert the threat. This sequence outlined in the EPPM has relevance for the terrorism preparedness campaign in the United States. The following sections use the EPPM framework to dissect and diagnose the terrorism preparedness campaign.

IS TERRORISM STILL CONSIDERED A THREAT?

EPPM relies on a threat to drive further action. One possible reason the terrorism preparedness campaign is failing could be that people do not view terrorism as a viable threat—it is not serious and they do not feel susceptible. Two sources of information are relevant to judging how U.S. citizens are evaluating the terrorism threat, media treatment of terrorist, and polling data related to terrorism. Combined, this evidence should indicate whether people still think of terrorism as a threat.

Media Treatment

The news coverage of an issue often follows a predictable cycle—what Downs (1972) calls the *issue-attention cycle*. The basic notion is that an issue bursts into the media and receives intense coverage that fades away before being resolved. The issue-attention cycle sequence is: (a) preproblem stage—a social problem exists, but has not garnered public attention; (b) alarmed discovery and euphoric enthusiasm—some dramatic event makes people suddenly aware of the problem and alarmed about its consequences; (c) realizing the cost of significant progress—people realize the cost of solving the problem is very high; (d) gradual decline of intense public interest—people get bored, discouraged, or feel threatened by the problem; and (e) postproblem stage—the problem moves to lesser attention or recurring spasmodic interest (Downs, 1972). Terrorism has been in the U.S. media for over 3 years now. It is logical to expect that interest has faded and the issue has entered the postproblem stage. However, once an issue is in the cycle, it can recapture public interest if issue-related stories have a strong news values and reemerge (McComas & Shanahan, 1999; Peters & Hogwood, 1985).

News values are the qualities reporters seek in a story. A common set of news values include: timeliness, event is recent; relevance or consequence, great potential impact on readers or viewers; proximity, news is local; prominence, it involves important public people; rarity, event is not routine; and human interest, interesting stories about average people (Treadwell & Treadwell, 2000). Terrorism has news values. Warnings about terrorism are given by government officials (prominence), can affect a large number of people, can be very serious (impact), and are dramatic. Every time the terrorism alert level changes in the United States, it is a national news story. Alerts issued by DHS are news stories as are statements of concern by politicians such as Vice President Cheney. People know the results of a terrorist attack will be serious. Chemical, biological, explosive, or radiological attacks will result in large numbers of deaths and serious injuries.

The terrorism issue remains in the public eye via the news media. However, the coverage focuses on the fear, not the efficacy of preparedness. In October 2004, a Lexis-Nexis search of U.S. news returned over 1,000 stories about terrorism in the United States. Only eight stories noted terrorism preparedness. Of those eight, only one mentioned specific preparedness actions. That story was about 40 hospitals in the St. Louis, Missouri, area that conducted a terror-related exercise (O'Neil, 2004). The other seven articles examined funding for bioterrorism legislation. The media are drawn to terrorism as *fear* not to *terrorism preparedness*. The news media focus on fear is further distorted through the nature of warnings provided by the federal government.

Most terror warnings are vague: There is some terrorist threat to some location or locations. U.S. citizens are advised to be more vigilant. The threat is ongoing. Sometimes the terror warning in the United States rises, sometimes it drops, but we are always on some form of alert for some reason. As Rudolph Giuliani noted, "You're safer knowing about the peril, even it makes people much more nervous" (Brady, 2004). His point is that the U.S. people need more specific information about the threats if they are going to be prepared. The federal government's policy, however, is to remain vague about the terrorist threats, often using security precaution as the rationale for not providing specifics. Skeptical people might view the vague threat as a means to keep the U.S. voters afraid and dependent on the federal government. Motives aside, the U.S. warning system seems better designed to induce fear than to provide strategies for coping with the threat. It is safe to conclude that the U.S. population perceives terrorism as severe.

As the issue-attention cycle would predict, scant media space is devoted to detailing solutions, such as preparedness measures, to the terrorist threat. The terrorist threat has news values, whereas preparedness measures do not. This is similar to the use of police drug bust stories during sweeps week, not stories about solutions to the drug problem. The issue has greater ratings potential than the solutions. The agenda-setting effect suggests that if the media are talking about terrorism, the public agenda should reflect terrorism concerns (Kim, Scheufele, & Shanahan, 2002; McCombs & Shaw, 1993).

Public Opinion Indicators

It is possible that U.S. citizens have become bored with terrorism and it is receding from their minds. If interest in terrorism has faded, so would a need for terrorism preparedness. The Kellogg Foundation study could reflect a lack of interest in terrorism, rather than a failure in the terrorism preparedness campaign. Public opinion polling data can determine whether that is the case. Polling data can be divided into two categories: (a) general issue polls designed to assess important issues in the United States, and (b) specific polls designed to assess terrorism perceptions.

A variety of organizations routinely assess what issues are important to U.S. citizens. During the 2004 presidential election, people were asked what issues were most important. These are general issue polls that locate the importance of terrorism relative to other issues in the United States. Throughout 2004, terrorism remained a salient issue in the United States. In February 2004, a Gallup Poll found 85% of the respondents indicated a candidate's position on the terrorism issue would be important to how they voted. A *Newsweek* poll taken 1 week later found 69% of respondents indicated terrorism would be important in deciding which presidential candidate they would support. In March, a CNN/*USA Today*/Gallup poll asked people what issue was more important—economic conditions or terrorism. Terrorism was selected by 26% of the respondents (Problems, 2004).

In June, a *NBC News/Wall Street Journal* poll found terrorism was the number two issue for voters at 22%, trailing only the war in Iraq at 25%. A September *FOX News/Opinion Dynamics* poll showed terrorism was the number two issue for voters at 22%, second only to the economy at 23%. A *Time* poll tracked which was the most important issue when deciding who a person might select for president. Terrorism remained second to the economy, which drew ratings of 23% to 25%. A *Newsweek* poll at the end of October 2004 asked registered voters, "Which one of the following issues will be MOST important in determining your vote for president of this year?" The number one issue was terrorism at 24%, followed by the situation in Iraq with 21% and the economy at 20% (Problems, 2004). Obviously the national elections could have had an influence in raising the salience of terrorism. The reason aside, the U.S. population has shown a consistent interest in terrorism in 2004, the same time span for which the *Redefining Readiness* study found they had largely ignored the terrorism preparedness campaign.

It is important to inspect beneath the veneer of the general issue polls to probe the depth of concern about terrorism. In 2004, the *Pew Research Center* and *Gallup* examined feelings about terrorism in the United States. The Pew study revealed that 20% of the U.S. people remain very worried and another 42% are somewhat worried there will be another terrorism attack in the United States (Pew Research Center, 2004). The Gallup poll indicated that 9% of the U.S. population are very worried and 25% are somewhat worried that they or someone in their family will be a victim of terrorism (Gallup, 2004). The *Redefining Readiness* report asked two questions related to the importance of terrorism. When asked how much they had thought about a possible future terrorism attack over the past 2 months, 39% said a little, 24% said a moderate amount, 6% said a lot, and 4% said a great deal. When asked how likely it was that a terrorist attack would happen in their community, 27% said moderately likely, 13% said very likely, and 5% said extremely likely (Redefining, 2004).

Specific terrorism polls have discovered that a significant portion of the U.S. population fears another terrorist attack in the United States. Terrorism is still viewed as a threat; there is severity and susceptibility for a significant portion of the U.S. population. However, the U.S. people seem to have adopted a defensive response, rather than a protective response. The terrorism preparedness messages do not seem to have been effective. People accept the threat, but do not seem to accept the efficacy of the terrorism preparedness messages.

Not only were people concerned about terrorism, they indicated a thirst for preparedness information. When asked how interested they were in learning more about government plans for terrorism preparation, both federal and community, 30% were moderately interested, 33% were very interested, and 14% were extremely interested. In addition, people felt the government was unaware of their concerns and the information they wanted about terrorism preparedness. A full 28% said the government cared or knew little, and 16% felt they cared or knew nothing at all (Redefining, 2004). U.S. citizens report they want information from the government about terrorism preparedness. Yet they are skeptical that the government understands their concerns or the information they desire. This would suggest there is a serious disconnect between the government's efforts to spread terrorism preparedness messages and the information that the public is receiving.

EFFICACY IN TERRORISM PREPAREDNESS

The *Redefining Readiness* report suggests that efficacy is problematic in the terrorism preparedness campaign. Clearly people cannot evaluate response efficacy or self-efficacy if they do not know the plan exists. Terrorism preparedness actually involves a set of plans—individual and governmental. Individual plans are actions people should take to prepare themselves for a possible attack. Governmental plans are actions the government orchestrates and recommends be taken in case of a terrorist attack. It is important to consider the efficacy evaluations for both plans.

Individual Plans

The focal point of the Ready.gov Web site is for individuals to prepare themselves and their families for a possible terrorist attack. The key message in the PSAs and the *Preparing Makes Sense. Get Ready Now* brochure is this statement from then Homeland Security Secretary Tom Ridge: "Terrorism forces us to make a choice. We can be afraid. We can be ready" (Preparing, 2003, p. 2). The U.S. government states it favors being prepared. The Ready.gov Web site provides detailed information on preparing a survival kit and making a family plan of action in case there is an attack. The material emphasizes how to do various tasks, with some information on why the task is important. The core point is that you are better off prepared than not prepared—an easy concept to grasp. The discussion of nose and mouth protection is another example of explaining why. A rationale is provided for why breathing protection is needed and how each of the various options works.

The Ad Council research suggests the Ready.gov Web site is successful in promoting individual-level plans. Over 2 million *Preparing Makes Sense. Get Ready Now* brochures have been downloaded from the Web site and 180,000 calls made to the toll-free number. Their research also found that people who have seen the PSA are more likely to take preparedness actions, such as creating the emergency kit or developing a family plan (Terrorism Preparedness, 2004). What limited evidence that does exist suggests that people are making favorable efficacy judgments about individual-level plans. People seem to believe the plans will be useful in a terrorist attack and believe they can be enacted.

Governmental Plans

The *Redefining Readiness* report indicates a serious failure to convey the efficacy of emergency preparedness plans for the governmental plans. Ready.gov is not designed to promote the efficacy of governmental plans.

The site does provide information about various types of terrorist threats, such as biological and chemical. The information explains symptoms involved with each threat and the advice to listen for additional information from government agencies in case of a terrorist attack. Efficacy information is lacking at the site. Detailed governmental plans cannot be given because there are a variety of contingencies. However, a basic outline of the general response to each type of terrorist threat could be provided along with a rationale for why that approach should be effective. The vagueness is akin to the maligned color warning system. People simply do not know what they should do at each color level or why.

One could argue that the Ready.gov site is not the best mechanism for presenting governmental planning information. State and community disaster planners could be considered more viable sources of information for terrorism preparedness because they are closer to the target audience. The Ready.gov site has a connection on its home page to a list of state emergency agencies. Using Illinois as an example, the state sites reinforce the federal message of being prepared. The Illinois site notes, "In any type of disaster, lives can be saved if people are prepared for the emergency and know what actions to take when it occurs" (Surviving Disasters, 2004, p. 1). Once more there is an emphasis on individual plans. Individuals are instructed to contact local emergency management officials to learn more about their community disaster plans.

The community plans should be a critical point for addressing the efficacy of governmental plans. People know they will experience the terrorist response initially through local officials who are coordinating with state and federal officials as necessary. It could be that the lack of knowledge and faith in governmental plans is derived from not knowing about and/or believing in the community plans. Whether community, state, or federal, it is doubtful that most people would know enough about the terrorism preparedness plans to make efficacy evaluations. By default, people then believe they continue to be at risk with no viable plan to protect them—one interpretation of the findings of the *Redefining Readiness* report. The government is as guilty as the news media in privileging risk over efficacy. More attention should be given to explaining why certain courses of action are necessary, (e.g., detailing why people should act immediately when the government recommends inoculations for small pox).

IMPLICATIONS FOR IMPROVING THE TERRORISM PREPAREDNESS CAMPAIGN

U.S. citizens continue to recognize terrorism as a risk. The news media and politicians offer reminders if attention was slipping away from terrorism. The news media and the government have established the fear elements necessary to move people to action. However, neither the government nor the news media have done much to promote the also necessary efficacy element. The implications will focus on the need to promote efficacy in order to increase the effectiveness of the U.S. terrorism preparation campaign.

The Efficacy Factor

The individual plans for terrorism preparation seem to be effective. The *Preparing Makes Sense. Get Ready Now* (2003) brochure begins with an easy-to-understand rationale for individual-level planning:

> The likelihood of you and your family surviving a house fire depends as much on having a working smoke detector and an exit strategy as on a well-trained fire department. The same is true for surviving a terrorist attack. We must have the tools and plans in place to make it on our own, at least for a period of time, no matter where we are when disaster strikes. Just like having a working smoke detector, preparing for the unexpected makes sense. (p. 3)

The Ad Council research suggests that people exposed to the terrorism preparation message take the desired actions. Buying duct tape and developing plans would suggest that people believe or hope it might work and can make these individual plans work. The problem resides with the governmental plans.

The discussion of the governmental plans involves federal and community-level plans. The *Redefining Readiness* report found that 74% of Americans were unaware of any government plans to handle small pox or dirty bombs. These data reflect a lack of knowledge about federal plans. On the community level, 55% felt their community had done little or no preparation for either small pox or a dirty bomb (Redefining, 2004). Citizens are not aware that the government has plans to cope with two likely terrorism attacks at any level. Using the EPPM, it is easy to understand why the terrorism preparedness campaign has failed. If people do not know there are plans, it makes little sense to think the plans have either response or self-efficacy. The target audience cannot make the necessary efficacy evaluations to promote a protective motivation response. Currently, the majority of Americans seem to be using the defensive motivation response manifest through avoidance or denial (Witte et al., 2001).

Possible Correctives

The terrorism preparedness campaign must redesign its approach to governmental plans. Two points are critical to the redesign: (a) providing specific information about governmental plans, and (b) promoting the efficacy of government plans. The governmental plans include both federal and community-level plans. Issues related to the federal level plans are examined first.

The federal government's general plans to handle either small pox or a dirty bomb are not top secret. Any person with the time and inclination can find information about how the federal government would address either of the two terrorist scenarios. I have heard both explained at public presentations. The Ready.gov Web site favors individual over governmental plans. Although individual plans are important to reducing the damage inflicted by a terrorist attack, the governmental plans should not be neglected. The Ready.gov Web site should be expanded. The information explaining each of the types of terrorist attacks should include outlines of the federal government's plans to be executed if such a terrorist attack would occur. Furthermore, there should be a rationale for why these plans should work (response efficacy) and their connection to the individual plans (self-efficacy). People should have a general idea of what the government will do, how it will help to lessen the damage from the terrorist attack, and what easy actions individuals can take as part of the response, such as getting vaccinated at a specified location. If people knew how the government planned to combat small pox, why the plan should work, and what they could do to help the plan, the percentage of people willing to follow the federal government's instructions should increase. Once people learn the information about the federal government's plans, the actual assessment of response efficacy and self-efficacy can be conducted to determine whether the messages are working or require refinement.

Community-level terrorism preparedness messages can be linked to state Web sites, but should use local community channels to get the message across to people. Local emergency management personnel need to become highly visible and vocal in the community. Local emergency managers should become known to speakers' bureaus and be active in community speaking engagements. They should be speaking at schools and civic and religious organizations to spread the word firsthand about the community plans for possible terrorist attacks. In addition, the local media should be pitched stories on a regular basis about the community plans. Local news media should be part of the preparedness drills. It is critical to have the news media cover drills to show community members what is being done and how it works. Efficacy should be promoted when people see, hear, and/or read about a successful local terrorism preparedness drill.

One promising finding from the Kellogg report was that 34% of respondents were very or extremely interested in personally helping with community terrorism preparation. The speaking engagements and media events could double as recruitment efforts to get more people involved. As people get involved, they learn the plans work (response efficacy) and become yet another apostle preaching the terrorism preparedness message. They also appreciate their role in the plan and their ability to make that plan work (self-efficacy). We can and should go beyond simply scaring people into buying duct tape and plastic.

CONCLUSION

Along the southern Atlantic coast, people know an evacuation order means a serious hurricane threat. Most people follow the evacuation routes and/or stay in designated shelters established by the government. Most people know and execute the government plan for hurricane evacuation. The evacuation order creates fear that a hurricane could hit. People evacuate and/or go to shelters because evidence proves it is the safest place; the plan has both response and self-efficacy. Warn people about a small pox outbreak and over 50% would choose to ignore the government's advice. They lack a context for interpreting the government's advice. They do not know why the action will help to reduce the threat. This resistant group has no faith in the plan because most do not know there is a plan. The 2003 rush on duct tape reflects the shallowness of America's preparation for a terrorist attack. We have created the fear, but have done little about efficacy. Fear sold the duct tape, but did people really know how to use it for protection or were they grasping at any possible sign of hope?

The government's terrorism preparedness message needs to expand to include efficacy content. More of an effort is needed to make people aware of governmental plans (at the federal and community levels), explain why the plans will work, and indicate what people will be expected to do to facilitate the success of these plans. Until the government expands beyond the message of fear and individual plans, little progress will be made. The news media do not have the motivation to promote plan efficacy—it simply is not attractive as a news story. It falls to the government to help people understand why its terrorism preparedness plans will work.

REFERENCES

Archibold, R. C. (2004, October 20). Cheney, invoking the specter of a nuclear attack, questions Kerry's strength. *New York Times*, p. A-20.

Brady, D. (2004, August 4). *Giuliani: Ignorance of terror isn't bliss.* Retrieved August 13, 2004, from http://www.businessweek.com.

Downs, A. (1972). Up and down with ecology: The "issue-attention cycle." *The Public Interest*, *29*, 39–50.

Gallup Poll August 9-11, 2004. (2004). Retrieved September 1, 2004, from http://www.pollingreport.com/terror.htm.

Kim, S.-H., Scheufele, D. A., & Shanahan, J. (2002). Think about it this way: Attribute agenda-setting function of the press and the public's evaluation of a local issue. *Journalism & Mass Communication Quarterly, 79,* 7–25.

Leventhal, H. (1970). Findings and theory in the study of fear communication. In L. Berkowitz (Ed.), *Advances in experimental social psychology* (Vol. 5, pp. 119–186). New York: Academic Press.

McComas, K., & Shanahan, J. (1999). Telling stories about global climate change: Measuring the impact of narratives on issue cycles. *Communication Research*, *26*, 30–57.

McCombs, M. E., & Shaw, D. L. (1993). The evolution of agenda setting research: Twenty-five years in the marketplace of ideas. *Journal of Communication, 43*, 58–67.

McMahan, S., Witte, K., & Meyer, J. (1998). The perception of risk messages regarding electromagnetic fields: Extending the extended parallel process model to an unknown risk. *Health Communication, 10*, 247–259.

O'Neil, T. (2004, October 29). Drill aimed to test terrorism response. *St. Louis Post-Dispatch*, p. B3.

Peters, B. G., & Hogwood, B. W. (1985). In search of the issue-attention cycle. *The Journal of Politics, 47*, 238–253.

Pew Research Center/Council on Foreign Affairs poll for August 5-10, 2004. (2004). Retrieved September 1, 2004, from http://www.pollingreport. com/terror.htm.

Preparing makes sense. Get ready now. (2003). Retrieved September 20, 2003, from www.ready.gov.

Problems and priorities. (2004). Retrieved November 1, 2004, from http://www.pollingreport.com/prioriti.htm.

Redefining readiness: Terrorism planning through the eyes of the public. (2004, September 14). Retrieved November 1, 2004, from www.cacsh. org.

Surviving disasters: A citizen's emergency handbook. (2003). Retrieved November 15, 2004, from http://www.state.il.us/ILHSASMnl.pdf.

Terrorism preparedness. (2004). Retrieved August 27, 2004, from http://www.adcouncil.org/research/impact_of_psas/.

Treadwell, D., & Treadwell, J. B. (2000). *Public relations writing: Principles in practice*. Boston: Allyn & Bacon.

Witte, K., Meyer, G., & Martell, D. (2001). *Effective health risk messages: A step-by-step guide*. Thousand Oaks, CA: Sage.

13

ETHICS OF COMMUNICATION

Scott R. Magee

J. Gregory Payne

Scott Ratzan

Although there are few bright line rules about ethics in any setting, we are able to reach two conclusions about ethical decision making in crisis communication. Ethical decision making by crisis communicators occurs twice: when they decide what to communicate and when they decide how to convey their messages. In both steps, history and experience provide the best guide for decision making. Absent the requisite experience, crisis communicators should first strive to eliminate the worst possible outcomes as they determine what to communicate. Past crises have taught us much about how to communicate the message correctly once the content of the message is determined. Those results are detailed here.

Uncertainty in the degree of risks to health and welfare is one of the key challenges to overcome in risk and crisis communication. Use of terms such as *infinitesimal*, *negligible*, and *probable* in defining risk, as part of crisis communication, places a responsibility on those who determine what information is valuable and salient for their respective audience(s). Moreover, once crisis communicators decide what message they are going to create, they need to find a way to communicate the message effectively in a way that does not exacerbate a crisis or lead people to misapprehend the seriousness of a crisis (either by over- or underemphasizing certain risks). A high ethical standard has thus become necessary for the purveyors of crisis communication, whether they are in government, academia, marketing, publishing, (new) media, or other fields of public address.

WHAT IS CRISIS COMMUNICATION?

Translating theoretical (imprecise and incomplete) and changing knowledge of causality and risk has developed into its own body of knowledge called *risk communication*. According to a U.S. National Research Council report, risk communication "emphasizes the process of exchanging information and opinion with the public" (as cited in Stern & Feinberg, 1996, p.___). No matter the nature of the crisis, whether it is a terrorist attack, a dangerous weather condition, or a health epidemic, those who fill the role of crisis communicator can benefit from studying incidents past to identify effective during crisis communication practices. "The objective remains the use of effective communication to avoid or avert a crisis in the first place, but if, for whatever reason crisis proportions have been reached, then that crisis must be handled- *managed*- until equilibrium can be restored" (Chamberlain, 1998, p.___).

The terms *crisis communication* and *risk communication* are inextricably linked and often used interchangeably. Crisis communication occurs when the risk is high to an audience that has experienced a breach in stability and in search of information that will help alleviate the problems at hand. *Crisis communication* has also been defined as the verbal, visual, and/or written interaction between an organization and its publics prior to, during, and after a negative event (Fearn-Banks, 2000). In both cases, communication is the conduit to define risk to a large group of concerned people, most of when are not experts on the specific concern.

FOUNDATIONAL UNDERSTANDING OF BEST ETHICS IN CRISIS COMMUNICATION

The way people behave in the realm of crisis communication can be varied and inconsistent. The variability arises for different reasons, including difference in risk assessment by the speaker and the listener, value judgments placed on possible benefits or losses that may result from a behavior, and outside constraints. Certainly this list is not exhaustive. Perhaps most important, the manner in which people communicate risk is shaped by amorphous concepts such as morality and ethics. At their root, morality and ethical decisions are shaped by other factors as well, some tangible, some not. In any event, the outcome of the decisions that a communicator and listener make results in tangible actions that are likely to affect the lives of the listener, the communicator, or third parties. The question then becomes, what should one do when engaging in risk and crisis communication? To find an answer to this question, we should examine some basic principles that exist in the background of any decision-making process. Specifically, we should ask whether there are overarching principles—philosophical or nonempirical—that provide a good guide from which we can extract scenario-specific rules. Even if there are not (and we submit that there are in fact no universal rules), we do benefit from this broader understanding. Particularly, this section outlines guidelines for the crisis communicator on how to determine what information a communicator is ethically compelled to disseminate to the public—those affected by the crisis. Regardless of what guidelines exist, a crisis communicator should rely first on empirical data and experience when they are available. In the following section, we present specific guidelines for how to ethically and effectively communicate the message once the communicator has decided on its content.

Although much has been written about various ethical approaches to societal, personal, and political choices, among others, certain processes are near universal. Jones et al. (1977) set forth four issues to consider when undertaking ethical deliberations, a prerequisite for an effective crisis communicator:

1. What are the facts? As simple as this may sound, there are few absolute rules that will apply to every situation all the time, and thus the nuances of a crisis become of paramount importance.
2. What are the ethical rules that govern a given situation? Sometimes these rules are explicit in a code of professional conduct, for instance. In crisis communication, these rules are not codified.
3. Are the principles (of Issue 2) applicable to the situation at hand?
4. What decision should one ultimately make? (Jones et al., 1977)

This chapter focuses primarily on Issue 2: What ethical rules should govern the conduct of a crisis communicator?

Social scientists have studied how the average citizen perceives risks, and the evidence suggests that the public tends to view risks differently than the experts (i.e., the scientific and policymaking communities). Although consumers do perceive risks differently than experts, this does not necessarily signify that *different* perspective translates into the *wrong* perspective. These two audiences receive different information, process it in unique ways, and respond to conclusions based on their own set of circumstances and concerns. Scientists and regulators, for a variety of reasons, tend to focus on measurable, quantifiable attributes of risks. The nonexperts, who are often the average consumers, receive different information from the media, whose proclivity is to focus less on the quantitative aspects of risks and more on the qualitative, value-laden attributes of risks—such as fairness and controllability. (This often translates into the public responding differently than the experts.) To examine the first principles, consider the following scenarios that involve some degree of decision making between a putative risk expert and a putatively less sophisticated group or individual. These scenarios purposefully demonstrate risks and crises that have different scopes (i.e., how long-lasting and widespread the crisis is) and imminence (how quickly the risks materialize into a crisis, if they have not already done so).

First, consider a doctor and a patient. The patient comes to an emergency room with a hip fracture. After a series of X-rays, CT scans, and physical examinations by a number of doctors, the patient is presented with the options of having surgery or letting the fracture heal on its own. Obviously, much of the risk in this case is guided by the doctor's expertise. Hopefully, the doctor will be able to offer the patient an intelligent opinion based on her experience in treating other patients with similar injuries. Even so, the patient may have other considerations in mind. How much will the procedure cost? How long will it take to recover? Am I more likely to recover fully if I have surgery? Will surgery put an onerous burden on my family? These are just a few of the questions the patient will have to answer, either with the help of the doctor or alone, before reaching an informed decision.

Thus, the doctor's advice is merely one element of the risk calculus. Knowing this, the doctor has the opportunity to skew the advice that she gives the patient based on how she anticipates the patient will answer the nonmedical questions. For instance, if the doctor feels strongly that the patient needs surgery, but the patient is resisting due to cost, what should the doctor do? One option is to emphasize the importance of surgery and minimize the risks of surgery in communicating with the patient so that the patient ultimately opts for surgery. The doctor thus overvalues the benefits of surgery intentionally to get a desired result. An alternative is for the doctor to dispassionately and objectively portray the individual risks and benefits of surgery and nonsurgery to the patient even if this results in what the doctor believes to be the wrong decision for the patient's overall health. In either case, the doctor, as a crisis communicator, is making a value determination that affects the patient's health.

Second, much like a doctor–patient relationship is the lawyer–client relationship. Here, imagine that a criminal defendant is meeting with his lawyer and they are faced with the question of whether the defendant should testify in his own defense at trial. Much like the doctor, a lawyer is restricted in his interactions with others to a certain degree by professional standards of ethics. The onus is on the client to make the final decision about whether to testify, albeit with help from the lawyer, who is of course our expert in this scenario. The lawyer in this instance is presumably experienced in trial work, understands how juries react to certain defendants and other witnesses, and so on. The lawyer cannot guarantee success to his client either way, and the client may have strong feelings about how compelling his testimony will be. Moreover, the risk borne by the client (his liberty) is much greater than that borne by the lawyer (success in a case). Thus, the need for a client to make a well-informed decision is paramount. The standards that a lawyer should use in deciding how to advise his client need to be clear. Two primary options arise for the lawyer: He could portray the risk that his client faces in a way that leads the client to the decision that the lawyer believes is the correct one, or he could portray the risks in a way that permits the client to subjectively determine the best course of action. The lawyer and client may not always reach opposite conclusions; however, in the cases that present the most difficult ethical issues in risk and crisis communication, the lawyer and client may disagree. For example, the lawyer may strongly believe that the client should not testify, but may have come to this conclusion for reasons that the client discounts. The lawyer is faced with the choice of communicating the risks in a way that leads the client not to testify by omitting certain facts, or of communicating all of the facts and risks objectively (or even persuasively) and in their entirety, after which the client decides that he should testify. Once the client makes the decision, the lawyer is ethically obligated to permit him to testify. Is there a point at which the lawyer's judgment about the outcome should trump his ethical duty to give a client all of the necessary considerations in reaching such a decision? If so, is there a way to determine exactly where and when this line should be drawn? For both the doctor and lawyer examples outlined earlier, the choice for the expert is whether ethical ends justify (questionably) unethical means.

The next two scenarios operate at the macrolevel, unlike the previous one-on-one crisis communication in the doctor and lawyer examples. A third scenario that presents ethical pitfalls in crisis communication focuses on a government agency disclosing a risk to the general public. For instance, does the color-coded terror alert warning system of the Department of Homeland Security (DHS) actually present a meaningful picture of risks that are encountered by the general public? The greatest pitfall in the terror alert system is that the descriptions of risks are too vague to be actionable, at least for the average person. Although there may be responses by local and state law enforcement to a change in the terror alert, merely disclosing a color to the general public does little to guide behavior. Even statements by government officials who say that people should be vigilant are not helpful. The government is then given the choice of disclosing specific reasons for the terror alert—something the government may not want to do for security reasons (e.g., disclosing information might inform would-be terrorist of information's source)—or offering specific tasks the public can do to improve its safety.

Yet even the latter can be detrimental. When the DHS suggested in February 2003 that people purchase duct tape and plastic to cover windows and doors to prevent injury from an anthrax attack, the action did not necessarily make people safer. It only makes people safer if the individuals buying the plastic and duct tape are at a greater-than-negligible risk of being subjected to anthrax and the individuals use the plastic and duct tape effectively. For those people who are not at a substantial risk of being subjected to an anthrax attack, purchasing and even using the materials properly would create a false sense of security. If the greatest risk to someone is not prevented by duct tape and plastic on the windows, he may be deceived into thinking that it is the greatest risk, and will thus be less concerned with more pertinent risks. What, then, does the government do when presenting acute risks such as terror alert warnings to make them meaningful? Only by communicating a risk in a meaningful way can the communication be ethical, but even this tends to beg the question. As in the lawyer example, the threshold determinations include deciding whether the communicator or the listener should make the value judgment as to the magnitude and likelihood of the risk.

Finally, let us consider crisis communication that occurs on an even larger level and for a more generalized harm. There is considerable dispute in the geopolitical community about the problem of global warming. Although the Kyoto Protocol on global warming became effective just recently, not all states are party to this treaty. States that have the strongest economic incentive to not be party to it—most notably China and the United States because they are the greatest emitters of greenhouse gases—decry global warming concerns as overstated and not adequately balanced against economic development. States that are party to the treaty, particularly some in Europe, formally treat global warming as a much more significant risk. This does not even take into account the nonstate interest groups that profess to be for or against Kyoto.

For the sake of this example, let us assume that the United States has a stable of experts who can credibly say that the global warming threat is overstated and the European community is being alarmist. Likewise, let us assume that the European community has a stable of equally qualified experts who have demonstrated through computer models that global warming is an impending and serious threat to the future of the environment and, consequently, the human race. Both camps are entitled to profess their views, but hopefully they would be constrained by ethical considerations. For the European community, to create an analog to the anthrax example discussed earlier, if individuals heeded the warnings that global warming is dangerous, they may alter their behavior, but they would likely not be lulled into a false sense of security simply by deciding to bike to work instead of driving. If a nation altered its behavior, say through legislative action, in response to the warnings of the European community, it might sacrifice certain economic benefits, but it is unlikely that the nation would then believe that it need not worry about global warming simply because it is taking proactive steps to address the real problem. Unlike the homeowner who is not at risk of an anthrax attack, but still buys duct tape and plastic, the nation reducing pollution emissions is addressing the specific problem that threatens it. Thus, even if we characterize the European approach as alarmist, such alarmism may be ethically acceptable because it does not create a false sense of security from the risk it seeks to prevent. Of course, it may present other risks, such as the risk of stalled economic growth that would then reach many parts of society. This is the exact argument of the Bush administration. Ultimately, when considering these two views then, the relevance of communicating information about possible dangers turns on a value judgment about the magnitude and likelihood of each risk.

As the foregoing examples illustrate, the need to implicitly or explicitly make value judgments in the course of communicating risks arises regularly. Unfortunately, values are not tangible items that we can count. In the literal sense, we cannot place six values on one side of the scale and four values on the other side and always come out with the right answer. Because value judgments are governed by ethical decision-making processes (that are fueled by crisis communication), we should strive to quantify, as best we can, the value judgments that we make. Philosophers, economists, and political and social scientists of all stripes have undertaken this task with respect to life decisions. Does their work provide any guide? Yes and no. As we see, different circumstances of crisis communication (i.e, micro vs. macro) require different approaches to valuation.

Seeger and Ulmer (2001) present two case studies in business ethics where CEOs of privately held corporations responded to fires at their factories. In both cases, the CEOs responded to the fire by declaring even before the fires had been extinguished that they would rebuild the factories and in the meantime provide health benefits and workers' salaries. They posit these scenarios as ethically appropriate under a system of virtue ethics, which is derived from personal character and "some fundamental tendency to act in an ethically commendable and admirable way" (Seeger & Ulmer, 2001, p. 371). As a matter of business ethics, the behavior of the CEOs was obviously ethically correct, but different contexts and situations (both businesses were privately owned with no shareholders) impact a widespread practice of such ethical stances.

The following discussion centers on first principles as a guide to valuing risks in crisis decision making. This task of valuing risks in a crisis is at its root an ethical decision. Each decision has the potential to help some people and hurt others. Moreover, crisis communicators do not necessarily know the outcome of the decision. They simply know what they believe are likely outcomes. Traditional studies of ethics supplement what we learn from experience to reach the *right* decision. Together, experience and less concrete ethical formulations lead to the necessary first principles that a crisis communicator should follow. We do not suggest that these principles be used in every situation. Often the crisis communicators are experienced enough in their field to not need them. Experience and empirical data are almost always a better barometer than more abstract principles. An ideal crisis, if there were such a thing, would be so similar to past crises that risk managers (policymakers and communicators) would be able to use successful techniques from the past. However, new crises and risks will invariably arise. This has been apparent to the entire world since September 11, 2001. As crisis and risk communicators are faced with new crises, these first principles will assist them in making ethical decisions. In the best cases, they will lead to the most effective course of action; at the very least, in all cases, they will make the communicator aware of when he or she is making a subjective, value-laden decision that may not have a single right or wrong answer. The two steps of an ethical decision-making process are to (a) make the right decision of what to communicate, and (b) communicate that information in the right way. We turn now to the first of those steps—determining the definition of *right*—by briefly testing several traditional theories on ethics and morality, and then we turn to the second later.

Immanuel Kant proposed one categorical imperative: "Act only on the maxim whereby thou canst at the same time will that it should become a universal law" (as cited in Newberry, 1999, p. 281). In other words, an individual should decide whether an action is good or bad (moral or immoral) not by introspection, but rather by considering what the result would be if everyone in the world took the same action. It is a results-oriented, pragmatic look at a theoretical question. Even so, it assumes that there is always one right or wrong answer to every question on morality or ethics. To apply it to the earlier four case studies exposes the flaws in using it universally in evaluating crisis communication. In the doctor–patient and lawyer–client scenarios, the communicator is operating at the microlevel in a one-on-one interaction. Thus, the result is that there will be success or not; the possible results are binary. To apply a Kantian theory to such a one-on-one interaction would be to give up our human instincts and treat all patients and criminal defendants as numbers, rather than as persons. From a practical standpoint, that is just unlikely. Moreover, if every doctor took the decision of surgery or not out of the patient's hands, or if every lawyer took the decision of whether to testify out of the hands of the client, the errors of judgment that doctors and lawyers have—and this is inevitable because not all lawyers or doctors are equally competent—would be magnified. With respect to the terror alert and global warming scenarios, Kant's method of making a value judgment fails because it becomes a tautology. Because the government officials in either case are making decisions that affect all, or for practical purposes all of their society, they put into practice Kant's hypothetical. The result is that the initial judgment about the prudence of a course of action by the government is neither reinforced nor debunked until the risk has either materialized or disappeared. Nothing is able to guide the value judgments of the officials before they act other than their instincts and knowledge.

Jeremy Bentham and John Rawls both advocated theories for the betterment of society that tried to create a better world for a large number of individuals. Bentham's utilitarianism and Rawls' maximin principle are best considered together. Utilitarianism strives for the greatest good for the greatest number of people. Rawls' theory emphasizes the benefit for the greatest number of people more than the greatest good, where we "rank alternatives by the worst possible outcome" (as cited in Rawls, 1999, p. 133). An example illustrates this idea. Consider the two following decisions and three different effects on people. Each decision affects the lives of each person, so that the benefit or loss of value to each person's livelihood is expressed in Table 13.1.

In Decision 1 the net benefit is 18, whereas in Decision 2 the net benefit is 23. Utilitarianism holds that Decision 2 is better because it nets the greatest good for the greatest number of people. Rawls' (1999) maximin theory argues that Decision 1 is better because everyone is benefiting, even if the net benefit from Decision 1 is less than that of Decision 2. This analysis is relevant to the study of crisis communication. The truly important part of this analysis is to show that, in a crisis situation, Rawls' approach is more relevant. Optimally, a crisis manager will respond to a crisis in a way that

TABLE 13.1.

DECISION	PERSON 1	PERSON 2	PERSON 3
1	6	8	4
2	11	25	9

gets everyone out of the crisis—a goal that has been memorably described as *getting the cow out of the ditch.*[1] Thus, a crisis manager can feel as though he is fulfilling his ethical obligations first by choosing to communicate those pieces of information and instructions that will help get *everyone* into the positive side of the column. Thus, for a crisis communicator, Rawls' maximin theory presents the right solution. The crisis communicator can feel comfortable in her ethical decision making if she tries to help everyone, rather than viewing some people as expendable. In this way, crisis communication is different from societal structure generally. We as a society accept some liabilities—such as the risk of automobile accidents—to gain some other benefit—such as speedier travel. However, in a crisis situation, we can eschew the risks because there is usually little benefit from taking on risks. More colloquially, there is nowhere to go but up. This is best illustrated by the terror alert example. Rather than give away information about a low risk to the nation as a whole, such as anthrax and the possibility of hoarding duct tape and plastic, it is more ethically appropriate not to take on the risk of a false sense of security.

If philosophical principles do not lead us to a correct and unified answer, perhaps a more traditional economic or quantitative approach will. We do not need to look too far to see that an economic approach to crisis communication faces the same problem as philosophical approaches—the need to make value judgments. Under this approach, the communicator would place a value on the (a) the probability that a risk will materialize, (b) the gravity of the resulting harm if the risk does materialize, and (c) the cost in dollars and lost life of taking certain actions to prevent the risk from materializing. In algebraic terms, we can view this as a balancing test, where the product of the probability P and the gravity of harm G is balanced against the cost C of actively preventing a risk (i.e., $PG \cdot C$. If $PG > C$, then a crisis manager should act, but if $PG < C$, he should not). In slightly more concrete terms, this formula is illustrated by the two different approaches taken to global warming in the example given earlier. The countries that have signed the Kyoto Protocol have decided that, in global warming, $PG > C$. Specifically, P is the likelihood of global warming occurring, G is the likely resulting harm (climate change, etc.), and C is the burden placed on coun-

[1] Ann Mulcahy, CEO of Xerox, says she always remembers the parable of the cow in the ditch. This came from an attendee of a customer breakfast, and he said, "When everything gets really complicated and you feel overwhelmed, he told me, think about it this way, you've got to do three things. First, get the cow out of the ditch. Second, find out how the cow got into the ditch. Third, make sure you do whatever it takes so the cow doesn't go into the ditch again." This is what she refers to, and passes on to the leadership team, whenever there is a problem. "The first thing is survival; the second thing is figure out what happened. Learn from those lessons, and make sure you've put a plan in place to recognize the signs and never get there again" ("The Best Advice," 2005).

tries and consumers in adjusting greenhouse gas emissions standards under Kyoto. The global warming example is apropos for this type of analysis because the cost is more easily quantifiable in dollars and cents than most. Thus, a crisis manager should have this analysis in his toolbox when costs are quantifiable.

Although we are unable to find in the foregoing examples concrete principles that are applicable in all cases, they do present considerations about balancing values of right and wrong that underlie the practical suggestions to which we now turn. An economic analysis is best where risks are quantifiable and the choices are binary. Rawls' (1999) maximin theory works best for crises that affect many people in a short period of time. Although there is no obvious choice for ethically deciding how to treat an individual (as in the doctor–patient or lawyer–client situations), a crisis communicator will rarely be in that situation; when they are, experience will be easier to come by because individual crises are more common than wide-ranging crises.

Practical Principles in Crisis Communication[2]

The second step to being an ethical crisis communicator is to effectively communicate the agreed-on message. Experience has taught us much about what in practice is most effective in crisis communication, from the perspective of both the communicator and listener. Because so many of the novel crises we are now confronted with, or expect to be confronted with, derive from health-related incidents, we can use health crises as concrete examples of what is most effective. Ultimately, effectively communicating risks and the most effective solutions to problems is arguably the single greatest goal in achieving an ethical program of crisis communication. This section focuses on the most effective strategies for communicating risks and solutions in crisis situations once the value judgments about what the substance of the communications should be has been made. Among other things, these principles provide the communicator with a way to gain credibility with the listener and minimize confusion on the part of the listener. The following six principles are identified in many literature sources:

1. Listening to and partnering with stakeholders.
2. Establishing, building, and maintaining trust.
3. Working with the media.
4. Developing the message.
5. Being honest, clear, and compassionate.
6. Being prepared.

[2]This section is a modified version of Ratzan & Melzer, (xxxx).

Each principle uses recent examples to illustrate the pitfalls and/or successes of crisis communication and the lessons that were learned in the aftermath of an event.

During 2001, several U.S. authorities were subject to an anthrax exposure. The actual risk of anthrax exposure to the general population was uncertain. A general assessment was that governmental communication resulted in inappropriate use of prophylactic antibiotics. As a consequence, there was a notable decline of public confidence in experts' assessment of risk and an unforeseen increase in panic within affected communities. The anthrax crisis of 2001 resulted in a plethora of research that provides communicators with a how-to of what and what not to say and do during such a crisis. More important, authorities generated their insights and drew from this experience. Scott Lillibridge, CDC Director of the Bioterrorism Preparedness and Response Program, stated, "We knew that communications would be important, but I don't think we knew it would be this dominant in the response" (*The New York Times*, 2001, p. ___). The United States could then apply some of the lessons learned to respond to new developing crises around Bovine Spongioform Encephalopathy (BSE) and Severe Acute Respiratory Syndrome (SARS) by early and frequent communication with multiple partners.

In Europe, there have been multiple crises, including acrylamide, the Belgian Coca-Cola scare involving Dioxin, and mad cow disease (BSE). Also, the SARS outbreak in southeast Asia and the avian flu's sporadic occurrence provide lessons. These events have not led to epidemics in Europe, but lay the groundwork for future successful crisis communication.

LISTENING TO AND PARTNERING WITH STAKEHOLDERS

In a crisis, the category of *stakeholder* expands to include anyone and everyone touched by the event. Embracing these groups as partners in the situation is an advantage in managing the reaction to the crisis. Some guidelines to identifying and involving these stakeholders include:

1. Seek to involve all parties that have an interest in the crisis at hand.
2. Demonstrate respect for those affected by decisions and include them early in the process.
3. Use a wide range of communication channels.
4. Recognize the fact that people will hold communicators ethically accountable. These partners may have a different view of the situation and listening may enable taking account of cultural and other variables.

5. Empathize with your audience and try to put yourself in their place.
6. Try to assess needs and public opinion through polls, surveys, interviews and other formative research.
7. Recognize cultural, economic, and political considerations that may complicate risk communication (Covello, 2003).

Partnering with credible sources—whether it is with the media, large corporations, trade associations, or politicians—can be a valuable tool to get a unified message out to the concerned public. For example, during the 1996 BSE crisis in the United Kingdom, there was a lack of strategies to elicit relationships with key factions. One result was that McDonalds decided to eliminate British beef from its UK menu and in effect set a policy for other countries and corporations to follow. Two days later, British beef was banned worldwide. A relationship that involved McDonald's may have been useful (Ratzan, 1998).

The appropriate level of stakeholder involvement depends on a number of factors. The Presidential/Congressional Commission on Risk Assessment and Risk Management (1997) offers the following factors to consider in determining the nature and extent of stakeholder involvement:

1. The complexity, uncertainty, impact, and level of controversy associated with the decision to be made;
2. The urgency with which the problem must be addressed; and
3. The extent to which participants can have a genuine influence on the decision.

A stakeholder analysis provides the decision maker with a profile of potential stakeholders for consideration in decision making and communication processes. Analysis includes the following information for each stakeholder group: needs, interests, issues and concerns, and underlying values; risk perceptions; level of interest and knowledge of the issue(s); knowledge gaps and misconceptions; and trusted information sources and communication preferences. The profile is verified and updated through dialogue with stakeholders throughout the risk management process.

Establishing, Building and Maintaining Trust

The process of involving and identifying stakeholders is just one step toward the ultimate goal of building trust and credibility. These two go hand in hand toward reducing fear and mistrust from interested parties during a crisis.

Research shows that if a low-trust source (e.g., a tabloid) criticizes a high trust source (e.g., the government), then the low-trust source will lose (Lofstedt, 2002). If a high-trust source supports a low-trust source, the likelihood for the low-trust source to win a crisis rises significantly. High-trust institutions receive more media attention than low-trust institutions. This puts certain conditions on how a high-trust institution can and should communicate to media and the public (Lofstedt, 2002).

Trust is, at its essence, a measure of how the public perceives competence and credibility (Shore, 2003). Each time a nation is faced with a crisis, the degree of trust should build and not erode. This also presents the caution to be honest and open as communication blunders have long-lasting effects. "Once lost, trust and credibility are almost impossible to regain. Never mislead the public by lying or failing to provide information that is important to their understanding of issues" (Covello & Allen, 1998, p. x).

During the anthrax crisis of 2001, the Centers for Disease Control (CDC) decided to issue a recommendation for postal workers to begin prophylactic antibiotic use of doxycycline. This was different from the earlier recommendation to the Capitol Hill workers to take prophylactic ciproflaxin. This decision outraged postal workers, who felt they were being treated less well than their executive counterparts. The CDC was aware of new information that doxycycline worked as effectively as ciproflaxin, but with less side effects. However, this information was not broadly communicated before the recommendation was issued. There has been a view that if the CDC had anticipated such a reaction and included postal workers in the discussion, this information could have been disseminated with the new recommendation (Vanderford, 2003).

Following the BSE crisis of 1996, a survey conducted in the United Kingdom found that three fourths of people surveyed felt it was hard to know whether government advice was free of political pressure. The same survey found that the government was less trusted by the public than the food industry (Hughes, 1998).

In June 1999, 26 children from one school in Belgium reported illness after drinking bottled Coca-Cola. Hundreds of people contacted the National Poison Centre to report complaints after having drunk Coca-Cola beverages. Cases of poisoning were also reported from northern France. Ultimately the illness was attributed to mass sociogenic illness. This was the subject of intense media coverage. However, it was a full week before Coca-Cola addressed the public about possible health risks, possible causes of the problems, and the actions undertaken. Coca-Cola's delay in addressing the public and informing health authorities was perceived as a suspicious evasion. Government and health practitioners did not receive detailed information from the company until nearly 10 days after the crisis began. Because of a lack of information about what was happening and in an effort to shore up public confidence, the national health authorities banned the sale of all Coca-Cola products. This led to the largest product recall in Coca-Cola's history, involving the recall of about 100 million packages and a loss of around 99 million Euro (Verbeke & Kenhove, 2002).

Working With the Media

Although personal, face-to-face communications during a crisis may be more accurate and effective, the reality is that one of the most important nonmedical sources of health information are mass media news reports (Bunn, 1994; McCallum, Hammond, & Covello, 1991). The public may receive 85% of its health information from media sources (Wilkes, 1997). During the U.S. anthrax crisis, survey respondents were asked where they had gone for information, and the top choices were local TV and radio and cable and network news channels (Pollard, 2003). Therefore, the information that the media disseminates must be accurate and reliable to prevent any public misperceptions.

In Sweden, after acrylamide was identified as a possible carcinogen found at high levels in fried foods, from April 23 to May 16, the four leading Swedish papers carried 112 articles on the scare. During the first few days after the press conference headlines such as "Cancer Poison Found in Food," "Alarm About Cancer Poison in Common Foods," and "Potato Chips Can Give You Cancer" appeared on the front page. The acrylamide scare was a major evening news item, making the top news story 2 days in a row and receiving major coverage on morning and evening talk shows over the first 2 days (Lofstedt, 2003).

During the BSE crisis in the United Kingdom, the continuous media attention had a large, negative effect on consumer behavior (Verbeke et al., 1999). The question was raised as to whether the media hyped up or informed the public (Verbeke et al., 1999). Inaccuracies reported during that time heightened public fear, forced the government's hand to a severe reaction that had devastating economic repercussions, and led to public skepticism regarding the government's competence, which has lasted to this day (Dornbusch, 1998). An examination of the coverage of BSE in the *Financial Times* found that the "media filled a void and became the primary health disseminator, drawing conclusions regarding health risks; the media engendered fear over BSE; the media misinterpreted scientific data about BSE; and the government was portrayed as unreliable and incompetent in dealing with the BSE crisis" (Dornbusch, 1998, p. x).

Today's media have a need for constant information updates to fill 24-hour broadcasts. Crisis communicators need to be aware that if they do not supply information, the media will report what they have. There are effective ways to embrace the media and enlist them as partners in reporting critical information during a crisis:

1. Requesting that reporters disseminate specific pieces of information (Mebane, Temin, & Parvanta, 2003).
2. Communicating to news media the importance of attributing information to its source (Mebane et al., 2003).

3. Being accessible to reporters and aware of deadlines.
4. Preparing key messages and repeating them during press interactions.
5. Saying only what one wants the media to repeat.
6. Focusing on what is known and reassuring the media to find out what is not known.
7. Providing background materials about complex issues.

Developing the Message

When communicating to the public is necessary, developing and testing (if time permits) the messages is of great importance. Due to the emergent nature of crises, that is not typically possible. Some preplanning and crisis communication preparedness may allow for some standards to follow when crafting messages.

The CDC usually develops messages for a variety of audiences; pretesting and audience research are standard steps in the development process. During the anthrax crisis, the CDC used communication monitoring as a proxy for audience input as the CDC continually shaped its messages throughout the crisis (Prue, Lackey, Swenarski, & Gantt, 2003). Because bioterrorism was new to the CDC, there were no existing messages and materials to use with nonscientific audiences; today there would be some material available. The CDC used communication inputs from field staff, media briefings, public inquiries, Web requests, and opinion polls to determine what the major messages needed to address. In the case of anthrax, messages regarding exposure and necessary precautions were needed to combat public fears. This method of message development had flaws and barriers to success, but the CDC was able to take some lessons that may prove useful in future crises.

BEING HONEST, CLEAR, AND COMPASSIONATE

There are many variables beyond the control of the policymaker or risk communication planner. One example of a common variable that predisposes individuals and multiple publics to vulnerability and poor health outcomes is literacy. Literacy disparities are of concern because low-literacy consumers/patients who do not understand health professionals' instructions are less apt to make the best decisions or receive good-quality care. To reach all people effectively, information must be provided in a variety of formats and reading levels. In calm times, this can be achieved with planning and forethought. In times of crisis, thoughts of public literacy are overshadowed by the need to release information quickly (Rudd, Comings, & Hyde, 2003).

What can be done in times of crisis? Including members of the intended audience when crafting the message is a possibility. A review process can include members of the lay public. Principles of plain-language communication and easy-to-read formats should be familiar to organizations prior to a crisis. Attention to literacy in relation to language, format, and structure of message will help broaden their reach (Rudd et al., 2003). Besides clarity, connecting with the audience with compassion, empathy, and reassurance are key elements to public communication in a crisis. "Even when the only message is uncertainty, admitting that 'we don't know' is a tried way to earn long-term public confidence" (Koplan, 2003, p. x). Communicating with understandable language, and refined and consistent messages, and communicating only what you know, are critical. Equally important is the person and method selected to put out the messages on behalf of an administration or organization.

BEING PREPARED

Based on experiences with crises in Belgium, Anthonissen (2001) outlined key elements of crisis communication: background information, details about the incident, actions undertaken, sympathy, and consumer reassurance. If these elements are addressed quickly, there is a better chance of avoiding a large-scale scare. Consumer reassurance, however, hardly can be provided immediately because it will only be successful when based on scientific evidence resulting from careful investigation (Verbeke & Van Kenhove, 2002).

The events leading up to the Swedish acrylamide scare offer an example for planned risk communication (Lofstedt, 2003). The sentinel event in the Swedish acrylamide scare was a planned press conference. Swedish National Food Administration (SNFA) was aware of a soon-to-be-published study conducted by researchers at Stockholm University (SU), and they jointly decided to head off a scare by holding a press conference on the subject. They sent out a carefully worded invitation to the press to attend the briefing. A press conference was selected as the best communication tool to deliver their message because they felt it would circumvent any unnecessary miscommunications with journalists and members of the public. In fact, SFNA had only held one press conference in the past 10 years, and the researchers had not participated in a press conference, nor did they have media training. The end result was a press conference that may have created the scare rather than prevented one (Lofstedt, 2003). What began as a well-planned event ended up providing media and communication experts fodder for criticism. The cryptic press invitation, coupled with a significant delay before the press conference, led to information vacuums, rumors, and speculation. Although the decision to hold a press conference was made carefully, key questions regarding the purpose of the event were not asked, nor was thorough analysis of available alternatives undertaken (Lofstedt, 2003). In ideal circumstances, both the press invitation as well as the press conference should have been pretested (Lofstedt, 2003). "The most kind thing one can say is that one handled it this way because of a lack of knowledge, that one did not realize what signals one sent out when it was done this way" (Hedlund & Larmet, 2002, p. x).

The six principles presented here summarize the lessons learned by past communication efforts during crises. Acrylamide presents the pitfalls of uncertainty about message and the need to cultivate relationships with the media. BSE demonstrates the necessity of involving stakeholders at an early stage and the importance of maintaining credibility as an organization. The Coca-Cola dioxin crisis illustrates the need for fast provision of adequate, reliable, and sufficient information to consumers (Verbeke & Kenhove, 2002). Anthrax is a prime example of the importance of establishing trust and the necessity of constantly monitoring the public's reaction to the crisis. September 11 taught the United States and the world that a crisis can occur without warning, and the need to communicate with the public will be instantaneous. Planning for these communications is the overarching lesson.

It may be simple to find flaws in past crisis communications. Those in charge worked under pressure, often without full knowledge of the facts, and did their best to communicate what they knew. Those responsible may take from the past lessons and results and build them into planning for future crises.

CONCLUSION

It is impossible to subtract the substance and methods of decision making in crisis situations and leave a shell of ethics. For that reason, it is impossible to create a shell of ethical rules that is universally applicable in crisis communication, as with any other aspect of life that by definition exists because its substance is unknown. However, one can conclusively say that good ethics demand that a crisis communicator communicate the information that is necessary and useful, and that he or she communicate such information in a way that it is readily understood by the listener. Beyond that, lessons learned through experience and first principles of right and wrong go a long way.

REFERENCES

Anthonissen, P. F. (2001). *Murphy was een optimist: Hoe ondernemingen door crisis-communicatie in leven bleven* [Murphy was an optimist: How companies survived thanks to crisis communication]. Tielt, Belgium: Uitgeverij Lannoo.

The best advice I ever got. (2005, March 21). *Fortune Magazine*.

Bunn, M. (1994). Aspects of consumer search for health information: A framework for structuring information problems. *Health Marketing Quarterly*, *11*, 75–98.

Chamberlain, M. (1998) *Avoiding, averting and managing crisis: A checklist for the future*. In S. Ratzan (Ed.), *The mad cow crisis: Health and the public good* (pp. 169–175). New York: New York University Press.

Covello, V., & Allen, F. (1998). *Seven cardinal rules of risk communication*. Washington, DC: U.S. Environmental Protection Agency, Office of Policy Analysis.

Covello, V., Peters, R. G., Wojtecki, J. G., & Hyde, R. C. (2001). Risk communication, the West Nile virus epidemic, and bioterrorism: Responding to the communication challenges posed by the intentional or unintentional release of a pathogen in an urban setting. *Journal of Urban Health: Bulletin of the New York Academy of Medicine*, *78*, 382–391.

Covello, V. (2003). Best practices in public health risk and crisis communication. *Journal of Health Communication*, *8*(1S), 5–8.

Dornbusch, D. (1998) *An analysis of media coverage of the BSE crisis in Britain*. In S. Ratzan (Ed.), *The mad cow crisis: Health and the public good* (pp. 138–152). New York: New York University Press.

Fearn-Banks, K. (2000). Crisis communication: A review of some best practices. In R. Heath (Ed.), *Handbook of public relations* (pp. 479–485). Thousand Oaks, CA: Sage.

Hedlund, I., & Larmet A. (2002, April 25). The alarm is not serious. *Expressen*, p. 1.

Hughes, S. (1998) Protecting health: Can the U.K. do better? In S. Ratzan (Ed.), *The mad cow crisis: Health and the public good* (pp. 111–117). New York: New York University Press.

Jones, W. T., Sontag, F., Beckner, M. O., & Fogelin, R. J. (1977). *Approaches to ethics* (pp. 2–5). New York: McGraw-Hill.

Kant, I. (1999). *Fundamental principles of the metaphysics of morals*. Reproduced in P. A. Newberry, *Theories of ethics*. Mountain View, CA: Mayfield.

Koplan, J. P. (2003). Communication during public health emergencies. *Journal of Health Communication*, *8*(1S), 144–145.

Matishak, M., Matsaganis, M. D., Payne, J. G., & Schulte, S. K. (2002). *Psychological aspects of decision-making in the event of a radiological emergency*. CDC workshop.

Maxims for Effective communication on health and risk issues for the World Health Organization. (1998). Washington, DC: World Health Organization.

McCallum, D., Hammond, S., & Covello, V. (1991). Communication about environmental risks: How the public uses and perceives information sources. *Health Education Quarterly*, *18*, 349–361.

Mebane, F., Temin, S., & Parvanta, C. (2003). Communicating anthrax in 2001: A comparison of CDC information and print media accounts. *Journal of Health Communication, 8*(1S), 50–82.

Ministry of Public Health. (2000). *Het Coca-Cola incident juni 1999 in Belgie¨: Evaluatie van de gebeurtenissen, discussie, besluit en aanbevelingen* [The Coca-Cola crisis June 1999 in Belgium: Evaluation of the events, discussion, conclusion and recommendations]. Brussels: Ministry of Public Health, Health Council.

Mullin, S. (2003). The anthrax attacks in New York City: The "Giuliani Press Conference Model" and other communication strategies that helped. *Journal of Health Communication, 8*(1S), 15–16.

National Research Council. (1989). *Improving risk communication.* Washington DC: National Academy Press.

Pollard, W. E. (2003). Public perceptions of information sources concerning bioterrorism before and after anthrax attacks: An analysis of national survey data. *Journal of Health Communication, 8*(1S), 93–103.

Presidential/Congressional Commission on Risk Assessment and Risk Management. (1997). *Framework for environmental health risk management—Final report* (Vol. 1).

Prue, C., Lackey, C., Swenarski, L., & Gantt, J. (2003). Communication monitoring: Shaping CDC's emergency risk communication efforts. *Journal of Health Communication, 8*(1S), 35–49.

Ratzan, S. (1998). Strategies for attaining public health. In S. Ratzan (Ed.), *The mad cow crisis: Health and the public good* (pp. 182–208). New York: New York University Press.

Ratzan, S. C., & Melzer, W. (xxxx). State of the art in crisis communication: Past lessons and principles of practice. In M. Haider (Ed.), *Global public health communication: Challenges, perspectives and strategies* (pp. 321–348). Sudbury, MA: Jones and Bartlett.

Rawls, J. (1999). *Theory of justice* (rev. ed.). Cambridge, MA: The Belknap Press.

Rudd, R., Comings, J., & Hyde, J. (2003). Leave no one behind: Improving health and risk communication through attention to literacy. *Journal of Health Communication, 8*(1S), 104–115.

Schulte, S. K., & Payne, J. G. (xxxx). Health communication challenges of an anthrax vaccination program. In M. Haider (Ed.), *Global public health communication: Challenges, perspectives and strategies* (pp. 363–378). Sudbury, MA: Jones and Bartlett.

Seeger, M. W., & Ulmer, R. R. (2001). Virtuous responses to organizational crisis: Aaron Feuerstein and Milt Cole. *Journal of Business Ethics, 31*, 369–376.

Shore, D. (2003). Communicating in times of uncertainty: The need for trust. *Journal of Health Communication, 8*(1S), 13–14.

Stern, P. C., & Feinberg, H. V. (Eds.). (1996). *Understanding risk: Informing decisions in a democratic society.* Washington, DC: National Research Council, Committee on Risk Characterization.

Vanderford, M. (2003). Communication lessons learned in the emergency operations center during CDC's anthrax response: A commentary. *Journal of Health Communication, 8*(1S), 11–13.

Verbeke W., & Van Kenhove, P. (2002). Impact of emotional stability and attitude on consumption decisions under risk: The Coca-Cola crisis in Belgium. *Journal of Health Communication*, *7*(5), 455–472.

Verbeke, W., Viaene, J., & Guiot, O. (1999). Health communication and consumer behavior on meat in Belgium: From BSE until dioxin. *Journal of Health Communication*, *4*(4), 345–357.

U.S. Department of Health and Human Services. (2002). *Communicating in a crisis: Risk communication guidelines for public officials.* Washington, DC: Author.

Zorn, M., & Ratzan, S., (2000 October). *Health risk communication.* Bethesda, MD: National Library of Medicine. Available from http://www.nlm.nih.gov/pubs/resources.html#2000

14

BREAKING NEWS AND HEALTH CRISIS

Mike Lewis

Public health crises are typically accompanied by intense media coverage, usually as breaking news, and high levels of media usage by the public. Journalistic reportage can provide significant benefits to the public during times of high uncertainty. Persons in charge of making decisions affecting public health during a crisis must know how to utilize the news media as a tool for getting important information to all stakeholders in a timely fashion. Understanding journalism work rules and following guidelines for communicating with news media ultimately aid the public in coping with health crises.

People watching morning TV hear an urgent beat and see a "breaking news" graphic suddenly fill the TV screen. Then a reporter is on the screen, reporting live about an outbreak of a mystery illness that has hit a group home for the elderly. Viewers see emergency response workers donning white bioterrorism outfits before entering the quarantined home while emergency medical crews rush the elderly into a parade of ambulances for transport to local hospitals. One man is already dead, and 15 people are being transported to local hospitals.

After a couple of minutes, the breaking news story goes away and the TV screen returns to its regularly scheduled programming. However, it is only a temporary respite. The crisis returns that day with team coverage during the noon, 5:00, 6:00, and 11:00 p.m. newscasts. TV journalists are reporting live at the group home, inside a hospital emergency room, and out in the neighborhood, each telling what little they know or can surmise about the mystery illness. The practice known as *media cannibalism*, a form of intermedia agenda setting, forces all of the local TV stations to cover the story. Collectively, the reporters and producers show the frightening stories dozens of times. Radio and newspapers join the frenzy. By the second day, the threat spreads to a church because a group had visited the home to sing Christmas carols shortly before the outbreak. A local high school is also dragged into the drama because one of its students had been volunteering at the home before returning to the student population. An entire community is thus affected.

Much of this crisis frenzy could have been averted, at least as far as news coverage was concerned. Medical authorities essentially remained silent. State public health officials referred questions to local authorities. The county health director was unavailable except for a brief telephone interview late on the first day of coverage. No one in the health department was designated to act as a public information officer to answer reporters' questions. The absence of a media spokesperson is all the more surprising because the county, situated in one of the top media markets, is one of the wealthiest in the country. Without any facts, reporters simply parroted whatever they were hearing from people at the various scenes, such as ambulance attendants. For example, several reporters used the terms *outbreak* and *quarantine* in their newscasts even though health officials never labeled it an outbreak and did not issue quarantine orders. Police and fire personnel described flulike symptoms, leading some to speculate whether a widely reported shortage of flu vaccine played a role in the illness.

In the end, the mystery illness turned out to be anything but mysterious. Investigators identified it as a form of norovirus, similar to the Norwalk virus that has stricken passengers on several cruise ships (Wilkins, 2004). Health experts say it is a fairly common gastrointestinal virus, rarely fatal, and often found in places such as senior facilities at certain times of the year. The one death may or may not have been related to the virus or some other malady the elderly man was suffering. Had someone in authority taken a few minutes to tell reporters that, or even their suspicion, the story would have lost much of its news value on the first day. This scenario, played out in Troy, Michigan, is quite typical of how many sudden disease outbreaks are covered by local reporters and gives clues as to how a terrorist event or catastrophic disease outbreak would first be reported to a confused and frightened public.

In this chapter, I describe the local news process and how it relates to the coverage of breaking news and public health episodes. Specifically, I examine the concepts of crises and breaking news and competitive economic motivations that fuel news coverage. The chapter concludes with a series of suggestions for public health officials on managing media coverage during a crisis, with an example of how one CEO turned a disaster into a public relations victory.

BREAKING NEWS

> God forbid there should be a disaster, but if there is, I hope it's on my beat.
>
> (A reporter's creed)

We live in an age of breaking news on TV. *Breaking news* is a concept that news workers use both as a marketing ploy and a description of how news is to be produced. It is typically accompanied by a graphic visual such as "Fox News Alert" and urgent musical tones designed to draw eyes and ears to the screen for the freshest events journalists have spotted on the news horizon. These days viewers may tend to think of breaking news as something akin to the 9/11 terror attacks, described as the worst crisis in modern U.S. history. However, in practice, most breaking news stories turn out to be about events that prove routine in nature. The tendency of news organizations is to frame routine events as sensational because viewers attend to breaking news out of a fear of real disasters or crises.

A crisis is a surprising event involving high levels of threat and uncertainty (Seeger, Vennette, Ulmer, & Sellnow, 2002). One of the basic understandings of crises is that they affect people, at the least stealing attention from normal or routine issues. Contemporary social life brings a steady diet of mediated local, national, and international crises (Quarantelli, 1998). Examples of crises include natural disasters, such as Hurricane Katrina or the eruption of Mount St. Helens, and unnatural events, such as Pearl Harbor, the Exxon Valdez oil spill, and the Challenger and Columbia space shuttle disasters. Public health events that qualify as crises include the influenza pandemic of 1918 and the SARS outbreak of 2004. Crises "affect large numbers of people and are marked by sudden onset, uncertainty, and lack of control, emotional reactions, and threats to lives and property" (Perse, 2001, pp. 53–54).

When a national crisis suddenly arises, people feel threatened. "Public crises are natural or manmade events that pose an immediate and serious threat to the lives and property or to the peace of mind of many citizens" (Graber, 1989, p. 305). Behavior is altered, individually and socially. There are large, sudden changes in how people think about issues (Graber, 1988). "Crises may involve fundamental loss of understanding regarding both what is happening and what can be done" (Seeger et al., 2002, p. 55). The revelation of a crisis is quickly followed by uncertainty, as normal routines are disrupted and potential threats revealed (Seeger et al., 2002).

The people's need for information is at its highest during crises. People seek information to determine what they should do—and to assess their risk, restore their peace of mind, and make sense of the events (Covello & Johnson, 1987). Numerous studies have shown that when people search for information to reduce uncertainty in a crisis, the majority turn to mass media channels (DeFleur, 1987; Greenberg, 1964). Since the birth of mass communication with Gutenberg's invention of movable metal type in the mid-1400s, people have relied on media to import ideas and information from beyond their individual sensory limits. This information-collection function is fundamental to the media. Journalist Walter Lippman (1922) explained that, because people cannot be everywhere at once, the media function as a window through which one can gaze all around the globe. Today, most people rely on TV to provide what is essentially a second-hand reality, a view of the situation as seen and heard through electronic video cameras and microphones positioned far from the viewers. Research strongly indicates that the public turns primarily to radio and TV for crisis information (Seeger et al., 2002).

The term *breaking news* is almost always applied to crises, but it is also used to introduce to the audience a wide variety of emerging news events. Sometimes the events are important enough to warrant interrupting or breaking into regular programming on radio and TV. For centuries, newspaper and magazine reporters raced to be the first to report or break a big story in print. The introduction of radio and TV provided an immediacy that supplanted newspapers in the race for journalistic scoops. Timeliness increases the value of the information. This capacity for immediate communication is important in the sudden development of a crisis, as leaders and viewers may learn of the events simultaneously.

Breaking news has now become part of the fabric of and public fascination with electronic journalism. In many respects, the term as used in practice is somewhat of an oxymoron, a contradiction in terms, in that it implies some extraordinary event, but is routinely applied to events as ordinary as traffic accidents and celebrity arrests. TV news takes its audience for a whirlwind ride from the scene of one breaking story to the next, regardless of whether the events have any relevance for the majority of the audience or for a larger understanding of the story.

News executives have learned that audiences are attuned to breaking news and other big stories. The primary convention of commercial TV news organizations is to attract large numbers of viewers, which boosts ratings. Big stories lead to big ratings, which stations then use to attract more advertising dollars and increase profitability. The economic theory of news making argues that decisions within news organizations are based primarily on financial aspects of the coverage. McManus (1988) found that events are more likely to become news if they are relatively inexpensive to produce and if the demographics of the anticipated audience are likely to represent the viewers sought by advertisers. For example, a TV station based in a big city may regularly choose to cover crime in that city for the 11:00 p.m. news because it would cost too much money in overtime to send the news workers to suburban news scenes, and because ratings surveys consistently show the crime coverage draws the most viewers.

These economic factors help to explain the steady decline in foreign coverage and investigative news reports by many news organizations, which are accustomed to large profit margins. "The mainstream corporate media are largely subservient to corporate interests, follow the sensation of the moment, and rarely engage in the short of investigative journalism that was once the ideal and that now takes place large in the alternative sphere" (Kellner, 2005, p. 181). Although some areas of reporting have suffered, others have flourished. Health in particular has developed into a booming business for mass media. Virtually every newscast, newspaper, magazine, and news Web site includes health information. News managers have discovered that, as the shrinking mass audience fragments into narrow, more specialized niches, health is one subject that interests practically every demographic: men and women; young and old; low, middle, and high income; White, African American, and Asian American. It explains why news media focus so much attention on health stories. Health is a topic to which everyone in the audience can relate. Good or bad, we all have our health.

Large audiences turn to news media during crises, but news organizations may not profit immediately because, depending on the story, coverage may run uninterrupted by commercials. However, most news managers believe that providing strong, sustained coverage of a big story provides long-term rewards by winning over new members of the audience who may become loyal viewers. As a result, news workers constantly survey the world for the next big story. Breaking news is often framed in terms of the impact on the audience, thus creating the impression that the audience members must continue to watch for their own well-being. Framing a story as a health issue or focusing on the health implications of breaking news is a strategy designed to build larger audiences.

When a major news event is identified, TV news executives instantly dispatch electronic news-gathering teams—reporters, camera crews, helicopters, and transmission trucks—to transmit live images and sounds from the scene. One element that almost all breaking news stories have in common is dramatic visuals, or at least the promise of pictures to come as soon as news crews arrive at the scene with transmitting equipment, including satellite phones with video capability, that allow them to broadcast live from virtually any location on earth. While reporters and photographers rush to the scene, producers back in the newsroom are contacting officials and experts to appear on TV and attempt to explain the events. If those people are unavailable, reporters at the scene will grab victims or their relatives, rescue workers, witnesses, and anyone else who is willing to go on camera and talk about the situation, whether they are informed or simply concerned. Journalists consider it a bonus if their interview subjects are emotional. In some cases, when it may be difficult to locate subjects quickly, such as on 9/11, reporters resort to interviewing other reporters. Regular programming may be preempted by coverage of the crisis. The interruption may persist for days in a national or international crisis. Although the onset of a crisis is typically sudden, its endpoint may be dragged out over time, testing the endurance of those people affected by the situation. "A crisis may be acute, as in the life-saving cases; but a state of crisis can also become chronic. The same elements are present—great danger and no escape—but the time frame is entirely different, and there is no one critical turning point" (Bok, 1999, pp. 110–111).

The event does not even have to be a real threat to garner media air time. Consider the following story from *The Washington Post:*

> Western intelligence agencies concluded that there had been a choreographed attack on numerous cities by terrorists wielding smallpox pathogens.
>
> By mid-afternoon, health experts realized that millions of people worldwide would soon die agonizing deaths. . . . Within hours of the first victims being diagnosed with smallpox, riots developed on the Polish border with Germany, whose border guards barred entry to nearly everyone. The Germans had enough vaccine to protect all of their citizens, while the Poles had enough for only 5 percent of theirs. (Mintz, 2005, p. A12)

Fortunately, the mock scenario was simply a drill. However, the threat is real, which is why government agencies all over the world participate in such drills and why news media report on the activities. Whether the threat is bioterrorism, SARS, mad cow disease, West Nile virus, a toxic train wreck, a nuclear accident, or a simple intestinal virus, one thing is certain—members of the news media will be among the first to spot the threat. Journalists will quickly inform the rest of the world, even before all of the facts have been gathered. They will do so through the lens of *breaking news*. As health agencies are organizing their initial response, journalists may be already transmitting information. Much of the early information may be inaccurate because the first sources may be wrong or intentionally deceptive. Because of the uncertainty of the crisis situation, rumors often emerge, which are picked up and repeated in the media. No time is available for fact checking or independent confirmation. During breaking news, many of the journalistic conventions for accuracy and multiple sources are replaced by the need to be first. During the Hurricane Katrina reporting frenzy, rumors were often reported, only to be retracted later.

WORLDS COLLIDE

In a crisis, the authorities—government, business, or some other organization—attempt to exert control in order to quickly restore a sense of normalcy to the situation. However, news media operate beyond the direct control of governmental authorities due to the free press protections in the First Amendment to the U.S. Constitution. Initially, authorities may attempt to control release of all information. However, all too often news media learn of the event through some other channel.

The news tip may come from anywhere: a tip caller, a police scanner, an ambulance service, a hospital employee. A victim or victim's relative may call the media in the belief that calling attention to the event will force officials to be accountable. It does not really matter where the tip comes from because there are too many possibilities to control the information. Moreover, during a crisis, health officials must focus on matters more important than tracking down the source of a news leak. All that matters is that the news media *will* inevitably learn about the event from somewhere, often before some of the health authorities are even aware of the situation. Efforts to control a leak or limit disclosure are almost always fruitless and often backfire.

The fact that the nature of the situation is still uncertain does nothing to slow the news train. In fact, the uncertainty may add a certain drama to the story. Once a news outlet learns of the story, it becomes a race against their media competitors to see which station will air the story first. A station or network with a reputation for being first with big stories will typically attract the biggest share of the audience when something happens. The station with the biggest audience makes the most money from advertisers. So while medical workers are worrying about public health, journalists are concerned foremost with the story and, ultimately, the story's ratings and impact on the news organization's profit-loss statement. Public health workers are responsible for fulfilling their mandate as public service agencies. Journalists, who may speak of public service as individuals, are employed by for-profit organizations with a mandate from stockholders to make money. Their individual careers may hinge on how their stories fare against the competition.

The rules are simple: Get on the air first and stay on the air as long as necessary. TV reporters must get dramatic pictures and show those pictures as often as possible. If victims are involved, get them and/or their families on the air. Of course, get the people who can tell viewers what is happening and who to blame. A few facts will help, but too many may complicate the story and make it difficult to air. TV works best with simple story lines. The newspaper and Web reporters, however, will be interested in the details and all the background information that is available. All journalists are operating under one fundamental rule: beat the competition.

The structural emphasis on intermedia competition and devoting live continuous coverage to breaking news events may contribute to the transmission of inaccurate information. Election 2000 is the now-famous example of news media mistakes. "If news is a rough draft of history, the draft written on the night of November 7, 2000 was rougher than most" (Jamieson & Waldman, 2003, p. 74). Despite journalists' proclamations of more careful reporting, inherent dangers exist. "Given current news-production conventions in the United States, news must be produced quickly, leaving little opportunity for research and reflections. Speed and 'the scoop,' unfortunately, are prized more than depth and insight" (Graber, 1988, p. 265).

People watch tragedy on TV primarily for three reasons: to feel empathy for the victims, to feel satisfaction that their own lives are still intact, and to assess the event's effect on their lives. Reporters will race after victims and their families and friends to get at the emotional aspect of the story, and to humanize the story so other viewers will feel more compelled to watch. They will also be calling the experts because emotions can only carry the story so far without information to put the events into some context. Pictures of someone crying do not answer for viewers the main questions: What is happening, why, and what will happen next? This is a point at which professional public health communicators and subject matter experts may exert some control over the message. When a health crisis suddenly arises, experts are often caught off-guard along with everyone else. While they scramble to get a handle on the situation, the phone messages from reporters start piling up. Mistakes are born and rumors created. These are reported and multiply through media.

For example, consider the anthrax bioterrorism attacks and the accompanying blizzard of press coverage. The Centers for Disease Control and Prevention (CDC) was mentioned in tens of thousands of news stories about anthrax during the final 3 months of 2001. The stories were written despite what amounted to almost an information blackout imposed by government agencies. "Usually helpful press officers were stonewalling, government scientific experts were not being made available for interviews, and public officials were generally failing to make accurate health information available fast enough. This mismanagement of news harmed the public good, reporters said" (Thomas, 2003, p. 12). Other observers have noted that the CDC was simply overwhelmed with inquiries and did not have the capacity to respond. This paucity of information was subsequently attributed to a number of factors, including inadequate information, the novel situation, and confusion about who was in control. The result, however, was the perception of stonewalling.

When the anthrax attacks began soon after September 11, government agencies were told not to talk with journalists as federal officials tried to speak with one voice. However, one voice could not answer thousands of press inquiries that came in to CDC and other agencies. Two weeks passed, and most reporters' questions went unanswered, except by a battalion of other, self-proclaimed anthrax experts who leaped into the news void. Reporters defended their use of outsiders as sources. One reporter noted, "I believe I was on firm ground in seeking opinions from non-governmental experts during the crisis. I had no choice" (Bhatt, 2003, pp. 14-16). CDC eventually assigned a team of media relations specialists and senior scientists to handle the deluge. By then news media had already made mistakes, including erroneous reports about new cases of anthrax and how to test for anthrax exposure (Chess, Calia, & O'Neill, 2004). The overriding mistake that government officials made during the anthrax crisis was believing they could control the information, even if they could not control the media.

FEEDING THE MEDIA MONSTER

The days of one evening newscast and a morning newspaper are long gone. Modern news media and bloggers transmit journalistic accounts around the clock, every day of the year, on paper, on the airwaves, cables and satellites, and online. Insiders liken the news media to a giant beast with a never-ending appetite. One local police chief speaking at a law enforcement seminar urged his colleagues to cooperate with news media, calling it a monster than needs to be fed. People who do not feed the monster are more likely to become food for it, he noted. Ironically, the news monster later turned on that police chief when he was accused of rape. Those of us who work in news media prefer to see ourselves in a less threatening, more civilized light: We are simply humble public servants, modern-day town criers. It is just that we come with a fleet of satellite trucks and helicopters, cameras, a cacophony of questions and never-ending deadlines—plus a mandate to beat the competition—when something called *news* happens.

"Disasters are media events." So begins the CDC's manual on crisis and emergency risk communication (Centers for Disease Control and Prevention, 2002). The manual includes detailed advice for working with media during a crisis. It contains useful information for individuals and their organizations when they are suddenly thrust into what is rapidly developing into a media event involving a health crisis. After enduring scathing criticism from journalists over its handling of the anthrax attacks, CDC established phone banks to handle questions from health care providers, journalists, and worried citizens. The agency also set up a facility for broadcasting press conferences (Thomas, 2003). The next section summarizes and critiques the CDC recommendations for working with news media. In addition, insights from a reporter's perspective are added.

WORKING WITH THE MEDIA

One of the most fundamental pieces of advice offered by the CDC is to cooperate with the media. People depend on news media for information, especially during a crisis, when their survival may be at stake. Neither local, state, or national governments own a system for relaying detailed information quickly to citizens during an emergency. Although the Emergency Broadcasting System is maintained by the government, it too uses private broadcasters to disseminate information. Radio and TV stations are mostly privately owned businesses licensed by the federal government to use a certain frequency on the public airwaves. In return, broadcasters are required, among other things, to provide government information to the public during a crisis. Thus, in a local or national emergency, electronic news media are structurally situated as the only conduit for the flow of vital information between government and society. Because they can tell many more people what to do and what not to do much faster than could emergency management officials with bullhorns, the media should be used. There is simply no alternative to the media for reaching the public during a crisis.

A second fundamental recommendation of the CDC is to provide answers. Even before the parameters of the crisis are known, certain questions can be anticipated: What happened? What is happening now? What will happen next? Why did it happen? How many people are sick/hurt/dead/evacuated/rescued/at risk? What can others do to protect themselves and their families and friends? What are symptoms and treatments? How does the danger spread? Where can people find help? Who's in charge? Who is to blame? What opportunities exist for pictures/video to accompany the story? Crises are high uncertainty events where the answers to these questions are not always available. This does not, however, excuse officials from addressing these questions. Reporters will still ask questions and will still press for answers. Even if forced to say, "We don't know yet," crisis communicators must be prepared to address these questions.

Those two points summarize the primary objectives regarding media during times of crisis. Using the media to communicate information to the public is a simple strategy, yet necessary to reach the public. Cooperating with the media speeds the process of reaching the public and helps ensure accurate information. The following section reviews specific tactics that will help in achieving the media objectives of public health communicators during a crisis.

The CDC recommends that a crisis plan be available at a moment's notice because that is all the time available during a crisis. During normal times, the media should be invited in to tour emergency operations centers and review operating procedures. Plans should include a nearby media operations center, where press conferences and interviews will be conducted and press releases and background information disseminated. The media will try to go to the scene anyway, in vehicles and news helicopters, and it is helpful if officials are nearby. Experts should also be available and should be reminded that plain talk trumps technical jargon. Scientific information should be explained in ways that are not condescending. Ultimately, officials are talking *through the media* to the public, who are concerned but may not be educated about the nature of the emergency.

It is also important that schedules be adjusted to meet the news cycle. Although deadlines often become irrelevant during a crisis or public health emergency, a media schedule is still important. Instead of meeting deadlines for the 6:00 p.m. newscast, officials may be asked for updates throughout the day's live continuing coverage (and well into the night, too). We now live in a 24-7 news cycle, and the media dedicate nonstop coverage to big stories. The media should be given a schedule of when and where officials will meet with them. These meetings may be once a day or every 30 minutes depending on the situation. Ground rules should be established and followed. Fair play includes not giving scoops to favorite reporters while others are waiting for the next scheduled news briefing. Doing so may create an ugly scene for that next briefing.

During breaking news, updates should be frequent and include the freshest information. In the beginning, fill-in-the-blank press releases that can be easily updated are helpful. Releases may also be posted on Web sites so the public and press can both have the same information. Messages should be consistent from different levels and agencies. Background information should be prepared in advance and readily available.

The CDC reminds officials that, for press conferences, the media will need to park their vehicles and set up their cameras and tripods. Depending on the crisis, a number of large satellite news trucks may arrive. Some electrical outlets to power computers, camera lights, battery chargers, and cellular phones may be helpful. Restroom access, of course, is another factor. A little food and water once in a while will be appreciated by overworked, underfed news crews. TV needs pictures, so easels or some other system to display visuals, such as charts and graphs, may satisfy some of the need for visual images.

Among the practical advice offered by the CDC is for officials to take a few minutes ahead of time to brainstorm questions and answers. A little rehearsal will help anyone's delivery, especially during the stress of a crisis situation. Officials should always be as truthful as possible. Above all, they should not lie or distort facts for even cynical reporters have a soft spot for honesty and the public is unforgiving to those who lie and deceive during bad times. It is acceptable and appropriate to admit not knowing the answer to some question, as long as you can also say what you are doing or will do to get the answer. The temptation to lie may be great. As Bok (1999) observes,

> In the care of the sick and the dying, in courtroom practice, in every kind of selling and advocacy—wherever the opportunities for deception abound, rewards are high, and time for considering alternatives often short—the danger of the formation of deceptive habits is much greater than in other lines of work. The word "crisis" then becomes a sufficiently elastic term to suit every occasion for lies. (p. 121)

If safety or logistical concerns preclude a media center at the scene, officials should consider providing videotape and still photos for the media. One method involves setting up a press pool in which one videographer and one still photographer, along with a reporter or two, are taken to the scene. They will then share their pictures and notes with others back at the media center. Another method is to provide visual images through video news releases. Even if fresh pictures are not available, file footage or photos will help. These images may also be made available on a Web site.

Experienced journalists have learned to expect mistakes on both sides. Officials will make mistakes and so will the reporters, who tend to lower their standards of verification during breaking news coverage. They simply do not have as much time to double-check information, and they can always correct it later when the crisis subsides. Mistakes should be corrected quickly and officials should then move on. Reporters will initially be cooperative because by nature they will be concerned for the events in their community. This goodwill should not be depleted by officials nitpicking small mistakes or holding simmering grudges against reporters with whom they disagree. Finally, officials should keep their cool, stay professional, and follow up with the reporter and then, if necessary, the reporter's boss.

If the breaking news qualifies as a national story, local reporters may require special attention to ensure they get what they need. The national reporters have national sources of information and will pack their suitcases and leave as soon as the crisis shows signs of abating. The local reporters will stick around for days, weeks, months, and years of follow-up stories, if warranted. Local reporters may end up with a chip on their shoulders because they were snubbed when Katie Couric came to the scene of a crisis. Ms. Couric is unlikely to make a second appearance, whereas local reporters will continue to cover the story. Still public health officials or other emergency management personnel should not be surprised when the national media dominate coverage. They are the 800-pound gorillas even if they don't act like it. Local media accept it and will be feeding information up the chain, particularly a local TV station that is affiliated with the network news crew. However, locals and their deadlines should not be disregarded.

A final point offered by the CDC that has particular relevance to breaking news concerns cooperation and working *with* media, rather than issuing orders *to* the media. Journalists do not work for public health, emergency management, or local government. They are willing to listen to the official side of the story, but they will decide what to put on the air. It is, after all, their newspaper, their program, their Web site, and their audience. As time goes by, they will look for other views on the crisis as well as the official position, even views that may disagree with the official line.

CONCLUSION

The thick, black plume of smoke on the horizon looked bigger than any fire we had ever seen. As news workers for three decades, the photographer and I had covered many fires over the years—with too many victims. But it would have taken the smoke from all of those fires combined to look like this dark monstrosity filling the afternoon sky.

We drove at breakneck speed to get to the Ford Motor Company Rouge Plant, scene of the February 1, 1999, explosion, even cutting across a snowy median of Interstate 94 to get to the scene faster. We scrambled to set up our live truck in the cold drizzle, across from a phalanx of ambulances rushing into and out from the burning steel factory. A police source was telling me via cellphone that it looked bad inside—at least one person was dead and several badly injured. Suddenly, my news partner called out, "Hey, there's Junior."

I looked through the rain-splattered windshield and saw a man standing on the corner, all alone, looking as sad as if he was watching his own house burn. As we rushed up to him with camera and microphone, William Clay Ford, Jr., Chairman of the Ford Motor Co., was clearly shaken. Looking pale and stunned, he called it "the worst day of [his] life" and told me that everyone at Ford is "family." Then, he was gone, away from the news media, off to the visit the plant, the hospitals, and the home of the first worker killed to privately convey his sympathy and support.

The industrial catastrophe ultimately claimed five more lives as workers succumbed over time to their severe burns. Another 14 injured workers survived. Ford Motor Company was fined $1.5 million by the state of Michigan for numerous safety violations related to the accident and agreed to pay millions more for safety programs and scholarships. Ford admitted no fault, and state officials agreed not to seek criminal penalties despite evidence that Ford had ignored safety warnings (Warikoo, 1999). In short, the industrial accident could have turned into a public relations nightmare for Ford, especially in Dearborn, where the steel plant and many of its workers are located, along with Ford world headquarters. But that's not what happened, largely because of the CEO's actions.

> The Rouge explosion showed why a sense of ownership still mattered. Once Ford heard of the accident, he didn't wait for lawyers to calculate liabilities, the public relations officials to determine an appropriate corporate posture, or an entourage to accompany him to the scene. Ford simply bolted for the door and spent the day attempting to help victims and their families. . . . Ford has shown that being a man of the people is a whole lot easier if you happen to be a real person yourself. No chapter in the public relations handbook tells you how to do that. (Pepper, 1999, p. 1c)

Responding to a health crisis, whether it is an accident, outbreak, or some other calamity, is always difficult. Doing so in the glare of live news media would test anyone's nerve and skill. During those times, organizational concerns should take a backseat to helping people affected by the crisis. This includes getting news out about what is happening, what people should do, and what is being done to contain the situation. Helping people is a virtuous, ethical act. Often the actions taken in the immediate aftermath of a crisis can bolster a company's or agency's reputation by demonstrating its values and commitment. Ford won over public sentiment with a simple act of sincerity. During the chaos and uncertainty of a breaking news event, simply doing your best under difficult circumstances, while maintaining respect for victims, is all that anyone—the press or the public—can ask.

REFERENCES

Bhatt, S. (2003). Useful lessons from reporting the anthrax story: A journalist describes what happened and shares what he learned. *Nieman Reports*.

Bok, S. (1999). *Lying: Moral choice in public and private life.* New York: Vintage

Centers for Disease Control and Prevention. (2002). *Crisis and emergency risk communication.* Atlanta: Author.

Chess, C., Calia, J., & O'Neill, K. M. (2004). Communication triage: An anthrax case study. *Biosecurity and Bioterrorism: Biodefense Strategy, Practice, and Science, 2.*

Covello, V., & Johnson, B. (1987). *The social and cultural construction of risk.* Boston: Reidel.

Kellner, D. (2005). The media and election 2004. *Critical Studies in Media Communication, 22,* 178–186.

McManus, J. (1995). A market-based model of news production. *Communication Theory, 5.*

Mintz, J. (2005, January 15). Bioterrorism war game shows lack of readiness. *The Washington Post,* p. A12.

Pepper, J. (1999, February 7). Explosion reveals heart of Ford's new chairman. *The Detroit News,* p. 1C.

Seeger, M., Vennette, S., Ulmer, R., & Sellnow, T. (2002). Media use, information seeking and reported needs in post-crisis contents. In B. S. Greenberg (Ed.), *Communication and terrorism: Public and media responses to 9/11* (pp. 53–63). Cresskill, NJ: Hampton Press.

Thomas, P. (2003). The anthrax attacks: A journalist assesses what went wrong in coverage of this story. *Nieman Reports.*

Warikoo, N. (1999, October 12). Safety problems took lives, but penalties differ. *The Detroit Free Press.*

Wilkins, C. (2004, December 14). Norovirus to blame for nursing home illnesses. *The Daily Oakland Press.*

ABOUT THE AUTHORS

Robert J. Alvey is a nationally recognized crisis communication practitioner. Previously the crisis communications director for the Arkansas Department of Health, he currently works with the Federal Emergency Management Administration.

Deborah Ballard-Reisch (PhD in Interpersonal and Public Communication, Bowling Green State University, 1983) was Professor in the School of Public Health at the University of Nevada, Reno, at the time this study was conducted. She is currently the Kansas Health Foundation Distinguished Chair in Strategic Communication at the Elliott School of Communication, Wichita State University. Her areas of research interest include health communication, bioterrorism, and risk communication, women's health, communication in families coping with cancer, and client/ provider communication.

Scott Barrett holds a Masters of Business Administration (1993) from Fort Hays State University, and is completing his PhD in Communication at North Dakota State University. His research focus is exploring employee perceptions of risk in high reliability organizations (HRO) and the relationship between these perceptions and various types and levels of organizational commitment exhibited by HRO employees.

Shannon A. Bowen (PhD) is Assistant Professor at the University of Maryland. Her research interests include moral philosophy, public relations, and issues management. She has authored numerous book chapters and published her research in a variety of academic journals.

Kristen Clements-Nolle (PhD in Epidemiology; MPH in Behavioral Sciences, University of California, Berkeley, 2001, 1993) is Assistant Professor in the School of Public Health at the University of Nevada, Reno. Her areas of research include public health communication, bioterrorism, and risk communication, social epidemiology, including HIV/AIDS and

STD epidemiology and evaluation with hard to reach populations such as homeless, incarcerated, and transgendered individuals.

W. Timothy Coombs (PhD, Purdue University) is an Associate Professor in the Department of Communication Studies at Eastern Illinois University. His research lead to the development and testing of the Situational Crisis Communication Theory (SCCT). SCCT provides recommendations from how crisis managers should respond to crises by evaluating key elements of the crisis situation. He has published widely in the areas of crisis management and preparedness. His research includes the award-winning book *Ongoing Crisis Communication* and his new book *Code Red in the Boardroom: Crisis Management as Organizational DNA*. He was part of the Darden School of Management's Batten Institute's "Defining Leadership: A Forum to Discuss Crisis Leadership Competency" and has a chapter in the related *Executive Briefing on Crisis Leadership*.

Denise Gorsline is an Assistant Professor in Communication Studies at Minnesota State University Moorhead and is completing her dissertation at North Dakota State University. Her research interests concern organizational communication. Prior to beginning her doctoral work, she was a vice president for distribution in a fast food company.

Keith Michael Hearit (PhD, 1992, Purdue University) is a Professor of Communication and Dean of the Lee Honors College at Western Michigan University. He teaches courses in organizational communication, public relations, corporate advocacy and crisis management. His research focuses on non-commercial forms of external organizational communication by corporations, particularly in those instances when companies are accused of wrongdoing. In addition to his book, *Crisis Management by Apology: Corporate Response to Allegations of Wrongdoing*, he has published articles in the *Handbook of Public Relations*, *Communication Studies*, and *Public Relations Review*, as well as contributed chapters in a number of edited books.

Robert L. Heath (PhD) is Professor Emeritus of communication at the University of Houston. He has written extensively on communication theory, rhetorical theory, social movement theory, issues management, public relations, crisis communication, risk communication and terrorism. He has recently lectured and participated in national and international panels on bioterrorism and biotechnology.

Tia Jenkins (MPH, University of Nevada, Reno, 2004) is Southern Regional Mental Health Disaster Response Coordinator, Southern Nevada Adult Mental Health Services in Las Vegas, Nevada.

Jennifer Kordsmeier is a graduate of the Department of Speech Communication at the University of Arkansas at Little Rock.

Kyle Leathers (MA, University of Nevada, Reno, 2003) is a doctoral student in Communication at Wayne State University in Detroit. His current research focuses on crisis and risk communication during hurricanes.

Jaesub Lee (PhD) is an Associate Professor of Communication at the University of Houston. His research focuses on relationship development and maintenance, leadership communication, media and information processing, risk communication, and intercultural communication.

Mike Lewis has worked as a journalist for three decades in print and broadcast news, both in the U.S. and internationally. He has covered a wide range of breaking news stories at the local and national levels. He is a former Mike Wallace Investigative Journalism Fellow at the University of Michigan, and holds a doctorate in communication from Wayne State University. Lewis is currently Director of the Journalism Program at the University of Michigan-Flint.

Feng Li received her Master's Degree at the University of Houston in 2004. She specialized in public relations with an interest in crisis communication, public health, and China.

Min Liu (PhD, North Dakota State University) is an Assistant Professor in the Department of Communication at Southern Illinois University-Edwardsville. She received her MA in English Language and Education from Shanghai Teacher's University, Shanghai, China. Min Liu's research interests include Health Communication, Risk and Crisis Communication, and Intercultural Communication.

Agnes N. Lyonga is a doctoral student in the Department of Education at North Dakota State University. This chapter presented here presents data collected while she completed her master's degree in communication.

Scott R. Magee received his B.S. from the University of Wisconsin-Madison, his M.A. from Boston University, and his J.D. from Boston College Law School. He has written on subjects ranging from patent law and to the First Amendment.

J.J. McIntyre earned a bachelors degree from Hastings College, a Masters Degree from South Dakota State University, and a PhD at North Dakota State University. He is currently an Assistant Professor at the University of

Central Arkansas where he is actively pursuing a research line in risk and crisis communication with specific interests in terrorism, crisis narratives, and chaos theory.

William E. Nganje is an Associate Professor at Arizona State University's Morrison School of Management and Agribusiness.

Julie M. Novak (PhD, North Dakota State University) is an Assistant Professor in the Department of Communication at Wayne State University in Detroit, Michigan. She is currently researching the relationships between risk communication, high reliability organizations, and democratic communication practices in the food industry. Her primary research interests include organizational, health, and risk communication. Previous to communication studies, she received an M.S. in international nutrition at Cornell University and worked in various public health and human service organizations.

Dr. J. Gregory Payne is an Associate Professor in Organizational and Political Communication at Emerson College in Boston and is an adjunct at Tufts University School of Medicine at the Global Health Department at Yale University. He is the co-founder with Dr. Scott Ratzan and Dr. Norman Sterans of the Emerson—Tufts Health Communication Program. He has served as a consultant to the Department of Defense, the CDC, U.S. State Department, U.S. Treasury, U.S. Treasury, City of Los Angeles and has been the editor of five volumes in the American Behavioral Scientist. He is on the advisory board of the *Journal of Health Communication, Media Ethics*, and *Communication Studies* and is the General Director of the Saudi—American Exchange, a grass roots, people to people diplomatic effort in the aftermath of the 9/11 tragedy.

Keri Pruitt (MPH, University of Nevada, Reno, 2005) is supervisor of the Kids to Seniors Korner Program, St. Mary's Mission Outreach, in Reno, Nevada.

Scott C. Ratzan (MD) is Vice President, Pharmaceuticals and Global Health, Government Affairs, Europe for Johnson & Johnson and Editor-In-Chief of the *Journal of Health Communication: International Perspectives.* Dr. Ratzan's prior position was as Senior Technical Adviser in the Bureau of Global Health at the United States Agency for International Development, (USAID), where he developed the global health communication strategy for U.S. funded efforts in 65 countries for the next decade. Dr. Ratzan has appeared on Good Morning America and Nightline as well as published articles in the *New York Times, Wall Street Journal* and *Financial Times* and in academic journals. His books include the *Mad Cow Crisis: Health and* *the Public Good, Attaining Global Health: Challenges and Opportunities*, and *AIDS: Effective Health Communication for the 90s*. He also has delivered many presentations including the Leiter lecture on Quality Health Communication for the National Library of Medicine and an address on risk communication for the National Cancer Institute that was selected in *Vital Speeches of the Day*. He was also founder and Director of the Emerson-Tufts Program in Health Communication, a joint master's degree program between Emerson College and Tufts University School of Medicine. Dr. Ratzan maintains faculty appointments at Tufts University School of Medicine and George Washington University Medical Center as well as the College of Europe in Belgium.

Barbara Reynolds has been with the CDC since 1991 where her communication expertise has been used in the planning or response to pandemic influenza, vaccine safety, emerging disease outbreaks, and bioterrorism. Internationally, she has acted as a crisis communication consultant on health issues for France, Hong Kong, Australia, Canada, former Soviet Union nations, NATO and the World Health Organization. She is the primary author of *Crisis and Emergency Risk Communication* (2002) and the CDC's Crisis and Emergency Risk Communication course, which is now taught in universities and other settings nationwide and internationally. In 2004, she launched a version of the Crisis and Emergency Risk Communication course for leaders. Her research and writings are focused on emergency risk communication best practices, integrating models of communication for public health, and building community hardiness in the face of disaster. In the past, she served as a senior press officer specializing in infectious diseases and vaccine safety issues, working with national and international investigative and science reporters. She is currently pursuing her doctoral degree in psychology.

Teresa Sacks (MPH, University of Nevada, Reno, 2005) is Project Director, Sanford Center for Aging, University of Nevada, Reno.

Matthew W. Seeger is Professor and Chair of the Department of Communication at Wayne State University. His research on crisis has appeared in *Management Communication Quarterly, Journal of Business Communication, Journal of Applied Communication Ethics, Journal of Business Ethics* and in several book chapters. He is the co-author of *Communication and Organizational Crisis* (Praeger, 20003) and author of ethics and *Organizational communication ethics: Decisions and Dilemma.* He is a co-principle investigator on the NSF funded Multi-Agency Jurisdictional Organized Response Project and has worked with the CDC on a number of issues related to crisis communication.

Timothy L. Sellnow (PhD, Wayne State University) is a Professor in the Department of Communication at the University of Kentucky. He has worked with the Centers for Disease Control and Prevention, the Department of Homeland Security Center for Food Safety and Defense, and the USDA on issues of risk and crisis communication. His work has appeared in *Communication Studies, Journal of Business Communication, Journal of Applied Communication Ethics, Journal of Business Ethics* among many others. He is the co-author of *Communication and Organizational Crisis* (Praeger, 2003) and is current editor of the *Journal of Applied Communication Research.*

Avinash Thombre (PhD, University of New Mexico, 2004) is an Assistant Professor in the Department of Speech Communication, University of Arkansas at Little Rock. His research interests center on analyzing the impact of new technologies and health communication. Presently he is involved in studying HIV/AIDS campaigns as well as perspective transformative experiences and its relation to health behavior change.

Robert R. Ulmer (PhD, Wayne State University, 1998) is Professor and Chair of the Department of Speech Communication at the University of Arkansas at Little Rock. He is co-author of *Effective Crisis Communication: Moving From Crisis to Opportunity* (Sage). In addition, he has published numerous articles that focus on how ethical communication, stakeholder relationships, leadership functions, and the discourse of renewal impact effective crisis communication. He is also an active practitioner and consultant on risk and crisis communication for public health departments, profit, and non-profit organizations.

Steven J. Venette is an Assistant Professor in the Department of Speech Communication at the University of Southern Mississippi. His research focuses on risk and crisis communication from an organizational perspective. He has worked with the Centers for Disease Control and Prevention, the National Center for Food Safety and Defense, and the USDA on issues of risk and crisis communication.

AUTHOR INDEX

SUBJECT INDEX

CPSIA information can be obtained
at www.ICGtesting.com
Printed in the USA
FFOW04n0154210116
20466FF

9 781572 737518